The Captain's Chair

The Captain's Chair

Jay Michael Brandow

To order additional copies of this book, contact:
Xlibris Corporation
1-888-795-4274
www.Xlibris.com
Orders@Xlibris.com
23375

Contents

Contents

This book is dedicated to the men and women lost at sea.

"Ready for the Storm"
by
Dougie MacLean

O' the waves crash in
The tide pulls out
It's an angry sea
But there is no doubt
That the lighthouse
Will keep shining out
To warn the lonely sailor

The lightening strikes
And the wind cuts cold
Through the sailors bones
To the sailors soul
Till there's nothing left
That he can hold
Except the rolling ocean

And I am ready for the storm
Yes Sir, ready
I am ready for the storm
I'm ready for the storm

Music and lyrics by Dougie MacLean
Published by Limetree Arts and Music
(PRS and MCPS UK)

Used with permission of Jennifer MacLean and
Dunkeld Records Ltd, Scotland.

Preface

The biting cold of the late November storm had torn away all sense of feeling from the flesh of shipwrecked Captain Walter Neal's fingers and toes, as he clung to what was left of his shattered boat. The steamer *Myron*, fueled by coal and made of white oak, had just been hammered to the bottom of Lake Superior, along with her sixteen-man crew. The pounding seas beat the last breaths of life from all hands on the tiny time-worn steamer, except one. The captain was the *Myron's* only survivor, and his time was running out fast. Mariners who know the odds of doing battle with zero temperatures, while at sea, say it would take a miracle for Captain Walter Neal, or anyone else to survive more than a few moments at best in the icy waters. Those who knew the "Iron Man" say he was too damned ornery to die. Just a mere mortal, but one short stroke ahead of the grim reaper. The doomed captain held fast to the floating pilot house, torn from the *Myron's* deck. The numbing cold continued to chew away at his severely bruised and battered arms and legs. Captain Neal had bent nearly every rule in the book during his lifetime and had a special talent for stringing cuss words together. Surviving the merciless beating of the icy seas would prove to be the biggest challenge he had yet to face in all of his 50-plus years.

The story is true, the characters are real, and while the story of Captain Walter Neal and the shipwrecked *Myron* have earned a few lines in the history books, this is the first time Captain Neal's forgotten personal story behind the headlines has been told beyond his family's kitchen table.

Many great discoveries have come about by accident; Louis Pasteur unlocked the secret of microbiology after his housekeeper lit a fire in the stove of his laboratory. While Pasteur believed he needed to perform his medical experiment in a chilled environment, he became outraged, upon returning home, discovering his housekeeper destroyed the "scientific control" he thought he had with his project. Rushing to place his experiment under the microscope to assess the damage caused by the heat, he was astonished to find that the exact results he was looking for came about because of his housekeeper's forgetfulness, "to leave his room alone."

This account of Captain Walter Neal and his family was also the result of an accidental discovery that came to light in bits and pieces over a 12-year period.

The "Captain's Chair" project claims to neither cure disease nor establish any scientific precedent; it does however, reaffirm that accidents do happen and we can often profit from them in some useful sort of way. This accident began as a simple search for a few paragraphs of history. As the restoration of this writer's rundown Victorian home was underway, it was thought a brief history of the house should be written and hung in the front foyer for posterity and the enjoyment of its visitors. That simple probe has resulted in the following pages of this book.

The discovery in this case, was an old snapshot that was found behind a plaster wall, during renovations. Four children were in the photograph, three boys and a girl, standing on the deck of a ship. As fate and luck would have it, the little girl, would later be identified; and located. She was, the last eyewitness to a part of history that nearly slipped away. Alexandra Stowe Johnston at nearly 100 years old, relives her past, and shares the story of her family and her beloved "Uncle Walter Neal." "Sis," as her friends have called her since she was three years old, also tells the story of her father, Alexander Johnston, who like "Uncle Walter" and his father William Neal, served as Great Lakes captains, in the days of wooden ships.

"The truth be told, Captain Johnston and his brother-in-law Captain Walter Neal were at odds with each other for most of their lives. Captain Johnston was a stickler for rules, routine, detail and proper decorum. Uncle Walter, on the other hand was quite the opposite as he spent most of his waking hours bending the rules, telling stories, and cussing," said his niece.

While many writers have written about ships, their captains and the sea, this effort takes us into the homes of these seafaring men and teaches us how a mariner's wife and children, and the rest of the Victorian world, dealt with the triumph and the heart-breaking tragedy that often accompanied such a noble occupation.

While written records and news reports of a single incident can often vary in content, every effort has been taken to insure accuracy and historic fact. What follows this preface is the result of a dozen years of research and face-to-face interviews with the last person to hear the voices of the characters in this book. Alexandra "Sis" Johnston, is in fact, the last eyewitness to a part of history that not only still touches us today, but has helped set the course in many cases for the future.

History, for far too many years, and for far too many former and current students, had come in the form of dreaded, drawn-out classes, taught by stuffy teachers, in even stuffier rooms, lecturing in torturous monotone voices, using dirty old books that had been in service, too many decades, too long.

The writer hopes this effort will shine a new light, for some, on the subject of history and bring with it the excitement of discovery, the charm of the bygone Victorian Era and the fun involved in visiting new places and meeting new people.

The words that follow this preface detail the lives, the loves and the losses of people much like ourselves. It is the story of who we were and where they, our ancestors, have taken us.

Chapter One

The Boy Sailor/Captain William Neal

England is where the story begins, in the early 1840's, aboard a wooden sailing ship crossing the Atlantic. Aboard is young William Neal, just a boy of ten years and the son of a sailor, who was also the son of a sailor, searching for opportunity and a new life thousands of miles from the place they once called home.

William Neal in his future years would touch many lives along the shores of the Great Lakes; his children and their children would become our neighbors, our classmates, our friends, and some, our extended family members. An ablebodied seaman and later ship captain, William Neal became known for his skill under the sail, his wit, his charm, his sense of humor, and his hard work. More importantly, William would also become known for his kindness toward his family and his shipboard crew.

For Captain William Neal there was no line of definition between his family and his crew. Often times, members of his immediate family were his crew. They all needed each other, they all depended on each other and eventually they would all pass through this life together, handing the torch to the younger generation of family mariners that would follow in their wake.

Captain William Neal / photo J. Brandow

William Neal began his sailing career as most young sailors do working long hours as a ship's second or third galley hand doing what he could to earn a few pennies. The boy sailor would soon learn the ropes from the veteran deck hands he was feeding and cleaning up after. Eventually, through this tutoring and hard work he would earn the position of deck hand to officer, and would earn the position of captain aboard his own ship. The ports of Detroit and Windsor are where William found his first work as a young seaman. A few years later he would move north to busy Port Huron on the eastern shore of Michigan's lower peninsula, in sight of Sarnia, Ontario.

The post Civil War lumber boom of the latter 1800's, would bring hundreds of sawmills to the region. Countless acres of old growth forest was being turned into lumber and lathe, that would need to be shipped to sea ports across the Midwest. There were jobs for everyone, it was this opportunity that lured Captain William Neal and his young family to the port of Bay City, Michigan. At the time, it was the third largest city in the state, because of its port, and one of the top lumber producing areas of the world. Any sailor who wanted to work could find a good paying job that would last the entire season. This is where Captain William Neal and his family, would call home, near the steamer crowded waters of the Saginaw Bay.

Mary Benton Neal / photo J. Brandow

Those who chose not to join the many fleets of the region's shipping companies as a sailor, could find work on dry land at the Davidson, Wheeler or other shipbuilding yards, at one of the saw mills, or other related businesses that lined the banks of the Saginaw River from its mouth up to Saginaw City a dozen or so miles upriver. These were boom times with a supply of lumber, work and jobs, that at the time seemed would never disappear. The port of Bay City was also the only place between Detroit and the Straits of Mackinaw where a wooden steamer of any length or tonnage could find a suitable dry dock for refitting or repairs.

The wealthy New York owners of the many saw mills had purchased thousands of acres of prime virgin forest lands where the tall white pines and the huge white oak trees brought them fortunes beyond belief. The "Lumber Barons" as they were called built huge mansions along the city's "Center Avenue," and in a few less populated areas on the edge of town. Quite a number of those urban castles that remain today have been restored to their original grandeur and elegance and still exude the demeanor of wealth and success. While the sun still shines through the stained glass windows of those palatial homes, the story behind many of the original families who occupied them has faded from memory and in many cases has been lost to oblivion.

With that information in mind the writer wishes to share the dream and the discoveries made while restoring one of those huge Victorian era homes before decay, neglect and time beckoned for the relief of the wrecking ball and the aid of the bulldozer. It was during this restoration effort that the story of Captain William Neal and his offspring was rediscovered, before it too, became foggy and forgotten never again to be shared or enjoyed.

The original endeavor was purely and simply to enjoy the hard work of restoring an old house, along with the thrill of adventure in searching for the many missing parts and fixtures that had either been lost, stolen or sold over the years. This writer found that the job actually became a front row window seat in a time machine.

It was also thought that it would be wise to chronicle the progressing phases of living the romantic dream of restoring an old house. Perhaps at the very least, the end product could have produced a small handbook of sorts, for lack of a better term, for those other daring dreamers who might also entertain similar thoughts of jumping off the deep end and beginning their own restoration project.

Surely those determined souls, who were just far enough off of their rockers to take on such a project, might spend a buck or two to read a home restoration pamphlet, written by someone who had been there, lumps, bumps and all. Perhaps the reader would even enjoy a few laughs and some of the lighter moments of the journey, while becoming familiar with some of the twists and turns their proposed project would bring them in the not too distant future. Those wise pamphlet buyers would also be briefed on the forthcoming budget busting challenges and the life's changes that would certainly come their way when they took the plunge.

Neophyte restoration artists could also become familiar with the stumbling blocks and other costly surprises of such a wallet squeezing adventure. These dreamers, although unlikely, might also find themselves regaining some of their common sense and be cautioned before hand about the hazards and pitfalls that would need to be endured, if they had actually hoped to chew and swallow the noble task they had bitten off. These adventurers are of a different sort and breed. Members of this proud but forsaken lot, can stand before the total wreck of a run down, beat up old house, tune out all common sense and the reality of their very presence, while envisioning the finished project right before their closed eyes. The dream is a familiar one, in full color and dimension, without a clue of how deep they will really have to reach into their pockets to make it all come true. It is this vision that often blinds even the most frugal of budget keepers. It is this enthusiasm that seduces even the strongest willed fixer upper into submission, to take the plunge, that will surely impact their lives for many years to come, if not forever.

It is this vision of renewal and majesty that will also, at some point, prompt those with "The Vision" to ask themselves repeatedly, "What was I thinking?" Sometimes before taking the leap of no return, willing visionists will often search for a good solid reason to hook their toes over the edge of their self-imposed challenge to secure a better starting foothold on that impending jump. Other visionists may explain their adventure seeking efforts as an avenue to preserve history, or they will claim that it's something they had always wished they could do and now is their chance. There may even be one or two, perhaps even three, who will admit they took on the project to keep themselves busy, to give themselves something to do, to stay close to home, or they just didn't know any better. The romantics will tell you it was their chance to reach out and touch a small piece of yesterday so it may be preserved for their children tomorrow. The psychologists will likely tell you, "those restoration dreamers are all off their rockers."

The single common factor shared by all of these artists is a rather unconventional and slightly bent sense of humor. This is a good thing; they will certainly need it to make it to the end and cross the finish line of their fix it up journey in one piece. But we must thank our lucky stars for their sense of humor, and for that momentary lapse of sanity as they signed the mortgage papers on that ramshackle castle. It is our great-grandchildren who will reap the benefits and the rewards. As a result of such a daring and dedicated action, a part of history and a part of each one of us, has been preserved. It's time to roll up your sleeves, empty your pockets, and lease the biggest dumpster that will fit in the yard; the fun is just beginning. What this visionist/dreamer both found and accidentally fell into was the long forgotten past. It is here, at this crossroads, where life as this writer once knew it changed forever and launched a new course, for life and career. I was heading willingly into the unknown, and almost certain financial ruin.

Was embarking on this project throwing caution and money to the wind, or was it a calling echoed by the voices of the past demanding the impossible? The undreamed of, but certainly welcomed discovery has not only helped turn the house into a real home again, it has become the journey of a lifetime; with more personal rewards than can fill a pirate's treasure chest.

The bodies of our ancestors rest in their graves, their places are marked in marble and granite, for eternity. They were who we are, they went to work to prepare the future. Just as each one of us, is working toward the future, and the futures of those who will follow in the foot steps we leave behind.

Now before this writer is a broken down old house that surely had finer days and many fond memories of growing families and the laughter they shared during those colorful days of times past. With the squeak of an old door hinge, or the creak of a wooden floor, one is left to wonder whose memories were they, and what happened to the voices now silenced by the years that have passed.

This is one man's account of the attempt to rebuild and restore a neglected time battered house, while at the same time stumbling over and collecting the tiny shards of tangible evidence of who once lived between its walls and under its deteriorating roof. This project is also about the many wonderful discoveries that came about during the journey, and more importantly, life as it unfolded for the people of the times, as told by the last eyewitness from the time history has dubbed the Victorian era.

A beat up old house, in a not so nice neighborhood, may not be the ideal project for the masses to take on but the results were certainly much better than anyone could have anticipated.

The writer shares an almost forgotten story that was buried for years in obscurity, a story that time tried to quietly rip away, had it not been for the results of a heavy crow bar, and the tip from a friend of a friend.

It took only one brick to hit me before I realized what was revealing itself and unraveling in front of me through the airborne plaster dust during the restoration project. It was that "same brick" that had struck repeatedly before the full impact of the discovery was realized and where it might lead. The story that was unveiling itself had been hidden for decades. It begins in this simple but stately Italianate Victorian 2-story brick house, not far from the Saginaw Bay and Lake Huron, but within earshot of the whistles of those ships still plying the waters of the Saginaw River. Steamships and the deep blue water of the Great Lakes lead this story's players to the region.

Bay City and Saginaw were once home to many of the states and the region's founding fathers, a presidential candidate, abolition leaders, war heroes, business leaders, entrepreneurs, and those of high society, not to mention those common folks who had found an honest way to make a few dollars to feed their families.

The people of Bay City and Saginaw were enjoying the boom times, while the "lumber barons" were building homes that looked like monuments to themselves. There were those business owners who were making their fortunes selling the building supplies used to erect those grand homes, simple frame houses and the shanties that many workers and their families called home.

Home of T. Riley Denison, supplier of building materials and supplies, the home was erected in 1872 and still stands today / photo J. Brandow

Working with the Bay County Historical Society, architectural historian Dale Wolicki shares his findings and details the list of former owners, of the Denison house. T. Riley Denison was one of those suppliers of building materials during the boom years, who did so well for himself that he could afford to build a home to rival those on Center Avenue, being built by the region's deep pocketed saw mill and shipyard owners.

The newspaper of the day, *The Bay City Daily Journal* reported that Mr. T. Riley Denison will build a new house on th Street, construction of Riley's Italianate Victorian home began in the second half of 1872.

Mr. Denison's voice and the voices of his family are a part of the past that surfaced thanks to Wolicki's research. Thomas Riley Denison was born at Cayuga, New York, on April 18, 1830. His father, the Reverend Avery Denison, brought the family to Michigan in 1832 and eventually settled at Troy in Oakland County. On August 23, 1857, Thomas married Jennie Flint and together they had six children. In November of 1865 Reverend Denison moved to Bay City with his wife and three sons, Charles H., Elias B., and Thomas R. Charles Denison was a noted local attorney. Elias Denison was involved in banking and real estate. Thomas Denison was involved in wholesale supply of grocers, hardware and merchandise to local merchants and companies, operating under the name of Delzell and Denison and later as Denison & Curry. Denison worked from an office, warehouse and dock at the Northwest corner of Water and Fourth streets. Denison eventually bought out his partners and operated under his own name.

Thomas and Jennie Denison resided at the 9th Street home for many years. On July 17, 1873, Thomas transferred the property to his wife. On March 3, 1886, Jennie Denison transferred the property to her brother-in-law Charles Denison.

There are no records or listings in the Death Index at the County Clerk's office; it appears that Thomas and Jennie moved out of Bay City in the Spring of 1886, writes Wolicki.

According to county records, on August the 12, 1887, Charles Denison sold the house on 9th Street to William L. Benham for five thousand dollars. At the time Benham was the Assistant General Freight Agent for the local division of the Michigan Central Railroad. Benham remained the owner of record until 1899. Benham's biography is minimal but noteworthy. Benham's profile was recorded in 1892, in the *Portrait and Biographical Record of Saginaw and Bay Counties, Michigan.*

> *William L. Benham, Our subject is assistant freight agent on the Michigan Central Railroad and is stationed at Bay City, having charge of the Third Division from Detroit to Mackinaw and from Jackson to Bay City. Mr. Benham was born in Ft. Atkinson, Jefferson County, Wisconsin and is the son of William H. and Lucy M.*

(Wright) Benham. His father was a native of Vermont where his grandfather, Silas, was a farmer and out subject's maternal grandsire built the first frame house in that part of Wisconsin where William L. was born. William H. Benham came West when twenty-one years old and engaged in farming and stock raising until a few years ago when he removed to Cedar Rapids, Neb., where he is now a successful ranchman. Our subject's mother is a native of Massachusetts and her father, William Wright, was a pioneer at Ft. Atkinson, Wis., where he devoted himself to farming. He was a devoted churchman of the Baptist persuasion, and at the time of his decease in 1861, was greatly mourned by the best people in the community.

Of a family of three children, our subject is the eldest. As his school days approached he was sent to the primary and grammar schools and finally finished at the Ft. Atkinson High School. He remained home until the age of fourteen years of age, when he began studying telegraphy at Oshkosh, and when fifteen years old was appointed operator at Fond du Lac in the Commercial office. Later he was with the Chicago and Northwestern Railroad at Oshkosh, spending one winter there as clerk and operator and was promoted to chief ticket agent. In 1875 he left the Northwest Road and located at Detroit, being chief clerk in the Commercial agent's office of the Michigan Central, and shortly afterward was made freight agent of the Michigan Central.

In October, 1886, Mr. Benham came to Bay City as assistant general freight agent of the division above mentioned, and he has now the charge and responsibility of the entire business as conducted from this point. He has a pleasant residence, which is located at No. 1009 Ninth Street at the corner of Farragut.

The domestic life of our subject is brightened by his wife, to whom he was married in Jackson. She was a Miss Mary L. Root and was born in Jackson. She was the mother of two children, whose names are Robert R. and Winwright. The family have been reared in the belief of the Presbyterian Church, of which they are consistent members. Politically Mr. Benham affiliates with the Republicans, believing the tenets of that party to be such as conduce most of the good to the general government. He is a member of the Michigan Republican Club.

There is no listing for Benham in the Death Index at the Bay County Clerk's office, He may have moved outside or died outside of Bay County, leaving no biographical information after 1899.

On July 31, 1899, Belle L. Bump became the owner of record of the 9th Street house. On July 20, 1905, Bump transferred the property to William J. Brand, who was a real estate speculator. Brand then sold the house to William O. Clift on April the 3rd of 1907. Clift was a local insurance agent also involved in real estate. On the same day he became owner of the 9th Street house, Clift and his wife Ella sold the property to Alexander Johnston and his wife Nettie for seventeen hundred dollars. Nettie was several months pregnant and would give birth to her only daughter, Alexandra, in the fall of 1907.

On May 11, 1931, Johnston sold the house to Josephine Covington and her mother Mary Lupien. Covington converted the building into apartments soon afterwards. In 1938, she moved into one of the units and resided at the 9th Street home until her death on November 6, 1953. In May of 1954, the Probate Court of Bay City transferred the property from the estate of Josephine Covington to Irene Warsaw per a July 12, 1949, purchase agreement.

Architecturally, the residence is an Italianate style structure. Popularized by noted architect and author Andrew Jackson Downing, the style is loosely based on the picturesque masonry villas of Tuscany. Ideally, prominent examples such as the Denison residence have a simple rectangular structure, low profile roof, and small entrance porches. They were built of brick with simple exterior wall surfaces to emphasize ornamental trim and features such as the paired brackets under the generous roof line. Windows were spaced equally across the facade with semi-circular or pointed hoods. Porches feature columns or posts and are trimmed in paired brackets that accentuate the entry. Attic windows, paired doors, elaborate chimneys, gable cornices with partial or complete returns, bay windows, and rooftop cupolas were popular details meant to enhance the picturesque image of the residence.

The old Denison house enjoys a rich historic past, and a short list of former owners that reads like a who's who of the region, but the grand old house had many dark years as well. The big brick house that once marked the edge of town had fallen victim to several of its former owners, after it was cut up into small apartments in the early 1930's. Penny-pinching landlords had refused to reinvest a dime of their profits into the once proud and well-deserving home. Following decades of neglect, the house that was once touted as one of the most modern homes of the region, boasting of working water closets on the first and second floors, had become a time-wracked wreck. The library with its tall windows had been partitioned off and became the living room of a small one-bedroom apartment. The main parlor with its oak mantled fireplace became another small unit, and the second floor bedrooms with their dressing rooms and walk-in closets became shabby studio units that rented for a few dollars a week. As time and the tenants took their toll, the cracked and broken plaster walls were covered with thick cheap wallpaper to hide the damage and make do through the tough times of the depression. Following World War II and the middle class migration to the suburbs, inner city homes and neighborhoods took a dive. Aging parents who began dying off in the late 1950's, 1960's and the 1970's were leaving their family homes to their children. Wishing to hang on to a piece of their childhood but not interested in living in their parents' homes, the new owners would put them up for rent. After all, collecting a handsome check on the first of each month, without making a cash investment seemed to make perfect sense, and would certainly allow for a few extras to make life in the suburbs the next best thing to heaven.

It didn't take long for the once moderately valuable homes to be destroyed by unconcerned renters. Following a costly eviction process, the houses were haphazardly patched up and placed back on the rental market where the cycle of destruction would continue.

With her darkest years now behind her, the once crumbling eyesore on Ninth Street could once again become one the finest houses in the neighborhood. Yet what good is a Victorian landmark if its history remains unknown? It was now time to assemble a few paragraphs of history and write a short tribute on paper. It would be framed and hung in the foyer where visitors could read it and enjoy a quick trip down a rather long memory lane.

With a little luck, maybe a photograph of the first owner, or at least maybe one of the house's early owners could accompany the brief history that was hung on the wall. A vintage photograph and a few paragraphs on the home's history, it was thought, would lend credence to the project and make a good conversation piece. The homemade historic tribute would serve as a link to the past for visitors who may happen by and help answer the repeated question of, "I wonder who lived here in the old days?"

Little did this writer know the quest to find that one photo and a few lines of history would become a mission that for a dozen years would consume every waking moment, and many sleeping ones as well.

This is a factual recounting of one old house and one American family with roots that reach deep into the past.

Several of ship Captain William Neal's children played an active role as the busy Great Lakes port of Bay City, Michigan, blossomed shipping many millions of board feet of lumber, to build homes and cities across the region, and beyond. It is Captain William Neal's son, Walter, who has become a part of Great Lakes history and lore, as the tragic sinking of his boat the *Myron* results in the loss of his entire crew coupled with his miraculous survival.

Captain William Neal's children
Nettie and Walter circa 1869 / photo J. Brandow

The oldest daughter of Captain William Neal, Nettie, would become an elementary school teacher and later marry a man under her father's command, a goodlooking, well-mannered Canadian by the name of Alexander Johnston.

In future years, it is Nettie and her husband Captain Alexander Johnston who will take up residence in the former home of T. Riley Denison, giving the family a proper captain's home with plenty of room for their growing family.

It was Captain William Neal and his wife Mary who taught their children to be self-sufficient, how to run a tight ship, and the value of a dollar. Mary Neal and her children would need to rely on those valuable lessons, to make it through the holidays in December of 1896, following the death of the family patriarch Captain William Neal.

Captain Neal's passing came as a total shock to those who knew him; it seemed that the hard working old school mariner with the tanned leather face would live forever. The slight man with the big beard died aboard his ship, the *Biwabic*, at his winter dock in Ashtabula, Ohio, as he was prepping his boat for the winter before heading home to be with his family for the holidays, and a well-deserved rest. Captain William Neal died at the age of 62 years.

The Biwabic / photo courtesy Ralph Roberts

Captain Neal's passing came too soon for the young English emigrant who came to the New World as a boy with his parents. Captain Neal was able to realize his dream of finding work and a good place to raise his growing family. Captain Neal also managed to preserve the family tradition of working as a merchant sailor while earning his way to the rank of captain of his own vessel. Captain Neal left behind a wife, a son and four daughters. While William Neal didn't live long enough to hold any of his grandchildren on his knee, his legacy has been passed down to them, so his story could be passed on to their children. His only son Walter Neal, by the time of his father's death, had achieved the rank of skipper of his own vessel.

It is the granddaughter of Captain William Neal who acts as our tour guide through the past. Her name is Alexandra "Sis" Johnston; she is the daughter of William Neal's first mate Alexander Johnston. It is Johnston who married his skipper's oldest daughter, Nettie.

It wasn't long before Alexander Johnston would find himself with a shipmaster's certificate and the skipper of his own command. It is the only daughter of Captain Johnston who takes us beyond the ship's records and past the yellowed and crumbling newspaper clippings of yesterday.

Nettie Johnston was the prettiest girl living in West Bay City and her new husband Alex was one of the best-looking men anywhere to be found in Bay City. Nettie loved and respected her father and it was her new husband who honed his skills under the command of her father. Captain Neal, from his photograph appears to be a slight man, with a head that seems much too large for the shoulders they rest on. It is the winter of 1896 where we begin, with the passing of the man loved by all who knew him.

West Bay City Tribune
Sunday December 6, 1896

Very Sad News; Captain Wm. Neal found dead in his boat.

> *Word was received in West Bay City last night from Ashtabula, Ohio saying that Captain William Neal had been found dead in his schooner. No other particulars given. Capt. Neal is resident of West Bay City, Michigan. He resided at 1207 Freemont Avenue. and had four daughters and one son. His son is Captain Walter Neal.*

Sis said, "Grandmother Neal spoke of her husband's passing on occasion. She would answer her grandchildren's questions about the grandfather they knew only by a photographic image, and the captain's key wind pocket watch, now carried by his widow."

News of Captain Neal's death spread quickly through sea towns across the Great Lakes, where the news was met with shock and disbelief. The captain was a very young man compared to the others who were still in command of their boats well into their later years. Merchant seamen were of a different way and cut from a different cloth. The newspaper boys had left their corners early after selling all of the papers they could carry the day word of the captain's passing was in print. Perhaps there was a misprint or a mistake on the name of the deceased mariner, some of the captain's friends asked. The undertaker had not yet received official word of any bodies being shipped to his embalming table. Surely the boys sucking suds, on Midland Street, at Emil Westover's Saloon, would know if news of the good captain's passing was true or just a sorry mistake.

Everyone in the merchant marine business knew everyone else; after all, the sailors at one time or another over the years had worked with each other on many of the same boats. A good number of them were related to each other by birth or marriage. Many, more often than not, had come within an arm's reach of the grim reaper weathering storm after storm, season after season; they were brothers of the sea, they were family. And like all good fun-loving brothers they would cherish the opportunity to enjoy each other, sometimes through the telling of tall tales, and sometimes through the fine art of fisticuffs. Every once in a while, after a few brews, the boys would engage in a little fun, just for the sheer enjoyment of the sport. But good or bad, love them or hate them, they were still brothers, and they would offer a salute while grieving at the passing of one of their own.

With the season at its end, and with many of the sailors already back home in both West Bay City and Bay City for the winter, there was plenty of "toasting" and "saluting" in the many saloons where the captain himself had bent his own elbow a time or two. One of the favorite watering holes for sailors and lumberjacks alike was Westover's Saloon.

While the pub itself still stands and looks much the same as it did more than a century ago, the name out front has changed over the years. The watering hole later became known as Larson and Rayman's. During Prohibition Larson and Rayman's manufactured cigars but took up its former trade of serving suds when the 18th Amendment to the United States Constitution was repealed in 1933. The tavern was again sold and became known as the Midland Street Stag Bar and like all good saloons, a respectable woman was not allowed inside; that's not to say women were never found sharing a smile or a drink with its patrons. The establishments most recent reincarnation came during the war years of the 1940's when the tavern became known as O'Hare and Sons, and a few years later became known as simply, O'Hare's. As of this writing, O'Hare's still serves cold beer and the biggest hamburger to be found on Midland Street. The ornate woodcarvings on the back bar at O'Hare's are the same carvings that were admired by those sailors and lumber jacks quenching their thirst more than a century ago. O'Hare's also plays an interesting role in the research of this work, and we'll take a closer look and explore the point, in the chapters to come.

History lovers will be pleased to learn that another time hidden Victorian treasure also still exists much as it did in the late 1800's. Bay City and West Bay City merged in 1905 and agreed to locate the new city hall on the East side of town. The meeting chambers of the West Bay City Council, and the community stage and theater that were abandoned at the turn of the century still exist, hidden from time, on the deserted second floor of a commercial building housing Wholesale Electrical Supply Company. The forgotten time capsule is located one block east of O'Hare's on the north side of the street. The milk glass entrance doors of the West Bay City Chamber still bear the old style lettering of the council's chambers, the community theater's front slanted stage and gas-fed stage lights still stand as if waiting for the season's next play to begin rehearsals.

The boys at Westover's Saloon hoisted yet another toast to the passing of their friend Captain William Neal. The next day's news would bring greater detail on the death of their friend.

West Bay City Tribune
Tuesday December 8, 1896

Captain Neal's funeral held this afternoon

The remains of Captain William Neal, whose sudden death by heart disease occurred Saturday morning in Ashtabula, Ohio reaching his home in this city at 10 O'clock Sunday night in charge of his brother, Captain W.D. Neal of Port Huron.

The funeral will take place from the family residence, 1207 Freemont Street this afternoon at 2 O'clock. The remains will be interred at Elm Lawn Cemetery.

The immediate cause of the Capt's death was heart disease. He was engaged in laying up his boat and worked within a half hour of his death.

He arose early Saturday morning, partook of a hearty breakfast, and went about to make the boat fast to the dock for winter.

About 9:30 O'clock he complained of a terrible pain in his chest, and his brother in law, Mr. Purser, steward of the boat, prepared and gave him a panacea.

He took the medicine, which seemed to relieve him for a short time, but a half hour afterward he again complained of pain in his chest and the steward started to prepare another dose of medicine.

While he was thus engaged Captain Neal, who was on deck, commenced to stagger, and Mr. Purser reached his side just as he was about to fall.

He assisted him to bed where he shortly afterward expired.

Captain William Neal was 62-years old at the time of his death, came to this country from England with his parents when a child. The family located in Detroit, and the son, when 14-years old, commenced sailing.

He came to West Bay City 22-years ago, locating to where his family now resides. He has been with Mills Transportation Company, of Port Huron, owners of the Biwabic for 23-years.

During his career the Captain visited nearly every port of importance on the chain of lakes, and was well and favorably known wherever he went.

The deceased leaves a wife, four daughters Misses Viola, May, Nettie and Anna who reside at the home and an only son, Capt. Walter Neal of this city.

Captain Neal was a member of Wenona council, Royal Arcanum in which he sured therein for $1,500.

Being both a leader and well-informed, Capt. Neal at least had the opportunity to get a glimpse of the new world to come. Sis recalls her grandmother saying, "Grandfather Neal would read everything he could get his hands on when it came to new inventions and discoveries."

A year before his death it was Guglielmo Marconi who sent and received the very first radio (wireless) signal, ushering in the age of radio. During the captain's lifetime he also saw the telephone begin to become a household fixture. In 1896, electricity had become old hat; it had already been serving the city of Bay City for the last 30 years.

Captain William Neal's commission, the *Biwabic*, was an old timer but a seaworthy workhorse with many good years still in her. Maritime historian Ralph Roberts says the bones of the *Biwabic* can still be seen today. After her service, she was stripped and turned into a barge as many aging vessels still are today. The *Biwabic*, after her tour as workhorse, was later abandoned not far from the Locks at the Soo (Sault Ste. Marie) near the Saint Mary's River where she reportedly lies with the bones of several other former proud wooden vessels. Bit by bit, she rots away with her wooden ribs of white oak disappearing with each season that passes and with each wave that touches her.

West Bay City Tribune
December 9, 1896

Laid to Rest

The funeral of the late Captain William Neal occurred from the residence 1207 Freemont Street yesterday afternoon. Rev. T.W. MacLean of Trinity Episcopal Church, Bay City, officiated at the services at 2 O'clock.

The floral offerings were numerous and very beautiful, among the special pieces being a boat and a pilot's wheel, both of which were comprised of roses.

The attendance was very large. The Pall Bearers were Captain George Smith, Captain George Lester, Captain F.H. Lennon, Captain William Patterson, William Dowdell and John Hutton.

Although the captain passed away more than 100 plus years ago, his family lives on. The captain's wife, Mary (Benton) Neal lived a good many years after her husband's death. From what has been gathered from the family stories passed down through the years, it seems Mary was never the same after her husband's passing. The photo that appears in this chapter is believed to be the last photograph taken of the captain and his wife Mary. The reader is advised to take note of Mary Neal's skeleton key pin that is seen on her collar; its discovery is a story in itself, we will explore later in this book.

One might not be able to tell from just a photograph alone, but truth be told Mary (Benton) Neal was a bit of a sourpuss. Or at least that's the word passed on by her family. Mary Neal spent most of her married life alone, taking care of a houseful of rather spirited children, while waiting for both her husband and his paychecks to find their way home. More than 20 years after the captain's passing, and up to Mary's death, life was less than pleasant for the merchant mariner's widow.

In the first decade and a half of the new century, Americans knew they would likely be pulled into the midst of a raging World War, and a deadly worldwide flu epidemic that would claim many tens of thousands of lives on both sides of the Atlantic. But it would not be the flu, consumption or whooping cough or any of the other forgotten diseases of the times that would claim the unhappy widow's life.

Mary had just finished a visit with her daughter Nettie, at her Ninth Street home. While shuffling her way up the block to wait for the streetcar at the corner, Mary took a terrible fall. Both the street and gutter were in poor repair and it was an accident waiting to happen, and it did; it would prove to be the beginning of the end for Mary Neal. A motorized ambulance was called to take the aging woman to Mercy Hospital. We must remember that the medical services we take for granted today did not exist in any form in 1918. Mercy Hospital was created by the Sisters of Mercy and was one of the city's first, and only a few blocks away.

Mary Neal's old bones were brittle, her hip was badly broken in her tumble and it would be only a matter of a few days before funeral arrangements would be necessary. It was said that Mary lacked a sense of humor and seldom smiled, but in light of her timing, and with her reputation for a pouting lip, one must ask the question, was it a parting joke or a coincidence that Mary's passing should take place on the 1st day of April?

Elm Lawn Cemetery / courtesy Leon Katzinger collection

Captain William Neal his wife Mary now rest together at Bay County's Elm Lawn Cemetery, along with their children, as do many of the region's former rich and famous.

The cemetery's grand stone entrance arch, the wrought iron gate and its guarding turret still remain today, time worn but as it was. For the shy or squeamish, we should remember and point out that grave markers and granite mausoleums were designed to commemorate and celebrate the lives and legacies of those whose innovations and hard work paid off and earned them both fame and fortune before passing on. In most cases, it was the dearly departed who chose the very stone that marks their final resting place.

The founding family of one of America's largest rental car companies also calls Elm Lawn its family's final resting place. If you would like to pay Captain William Neal or his family or any of the others a visit, you are certainly encouraged to do so. How many of us would like a visitor to stop and say hello once we too have been placed to rest for eternity? To visit Captain William Neal's family plot, enter Elm Lawn through the stone entrance arch.

Elm Lawn Chapel / photo J. Brandow

Once past the chapel you will turn left, on the first road you come to, travel about 30 feet and pull over to the side. On the right you will notice the Neals' weather worn white marble marker. It's topped by a sphere and is easy to locate. The "Neal" name will be facing you to the north. About 113 paces to the east you will find the grave marker of son Captain Walter Neal and his wife marked with a red granite stone, and to the south from where you stand, you will face the rear side of the gray granite marker of Captain Alexander Johnston and his wife, Nettie.

Sis said, "Captains William and Walter Neal, being of good humor, may appreciate a joke or two if the visitor cares to share one, and while we can respect their passing don't be afraid to chuckle or let out a laugh. If the roles were reversed, both William and Walter would certainly be trying to out do each other with a bit of wit and perhaps a belly laugh or two."

There are many other discoveries to be made at Elm Lawn; ship builder Davidson and fleet owner Boutell now call this stone garden, of eternity, home for their mortal remains.

The DeFoe family plot is also located at Elm Lawn. The DeFoes were the area's modern ship builders who began their modest company selling kits to build rowboats. Some of the DeFoe accomplishments include the Presidential yacht of John Fitzgerald Kennedy, along with Navy patrol craft used in World War II, Korea and Viet Nam. The Company closed it doors in the 1960's, after building several destroyer escorts for the Australian Navy. The DeFoe Company was also responsible for building a very special ocean going vessel; it is known by the name, the *Knorr*. This was the very vessel whose crew put an end to the mystery of the famed but elusive wreck of the *Titanic* that went down after striking an iceberg on its maiden voyage from England in 1912. The captain and crew of the *Knorr* discovered the remains of the famed wreck off the coast of New Foundland.

Visitors will also find the grave of Union Civil War General Benjamin Partridge. Partridge was a former Bay County Sheriff who mustered troops to fight in the Civil War. According to his records, Partridge was present for a great moment in American history; he was at the Courthouse at Appomattox and witnessed the official end of the Civil War. Partridge was only a few short steps away from Confederate General Robert E. Lee when he surrendered to Union General Ulysses S. Grant putting an end to the American Civil War.

Chapter Two

"Stranger in the house"

"Mother! Mother! who is that man in our house? Mother, who is that man, and why, is he living here?" asked Sis. Nettie (Neal) Johnston stood busy at the wood stove in the kitchen, slowly stirring her simmering beef stew. With her left hand resting on her aproned hip, an amused smile began to appear from the corner of Nettie's mouth. The twinkle in her hazel eyes began to sparkle, as she turned to answer her three-year-old daughter's questions. When she fixed her gaze on little Sis's eyes, Nettie could hardly hold back the smile from her powdered face.

"Young lady, that is your father!" Little Sis Johnston stood in the kitchen doorway and listened closely to her mother's words. Precocious and determined, Alexandra Stowe Johnston took a moment to try and realize what her mother had just told her, "my father? My father? Why is that man living here?" "Young lady," Nettie repeated, "that man is your father and you are his daughter. He is home for the winter and he'll go back to work on his ship." "On his ship? What kind of ship?"

Nettie raised her index finger to her lips and softly said, "shhhh, children should be seen and not heard." Not wanting to disturb "that man" and trying not to draw his attention, little Alexandra quietly returned to the front of the house and peeked into the library for another look at the big man with the mustache, working at the roll top desk. After a few moments of study

Alexandra slipped to the back of the house where her mother was preparing dinner. "There was no man living here before Mother!" Still smiling, the petite captain's wife sat her daughter at the kitchen table and served her a small dish of peaches. This would not only keep the little lady quiet for a few moments, it would give Mother some quiet time with her daughter to help acquaint her with the life of a typical mariner's family.

It was only mid-afternoon but it was already dark outside. It was late in the season; the leaves on the barren trees had fallen from their lofty summer perch months ago and the chill in the air made the snow crunch when you walked on it. Captain Alexander Johnston was home from another successful and profitable season and was now with his family, at least for a little while. Johnston was a well-established steamer captain on the Great Lakes and was known as a captain's captain for his attention to detail, his flawless pilot's record and his insistence on following the rules.

Nettie Neal / photo Anna Mae Wires

Captain Alexander Johnston was a big man with broad shoulders standing just over six-foot tall. Johnston was also perhaps one of the best-looking and best-dressed captain's on the Great Lakes. The captain preferred to wear a business suit rather than a normal captain's uniform while he worked, although he would don his captain's cap while at his perch above the wheelhouse. Johnston knew he was a handsome man and there were times, when the swaggering freshwater swashbuckler would surround himself in the air of conceit. The captain enjoyed the respect he was shown by members of the community, the good captain was proud of his good looks, his thick strawberry blonde mustache, and his station in life.

Alexander Johnston / photo J. Brandow

In his early years, the up and coming able bodied seaman found his way to meet the petite and beautiful daughter of his boss, whom he was sailing under as first mate. The title of ship's captain was held in high regard, and with it came the money and respect for such a notable position both socially and in the world of commerce. Many captains of the day were held in the same high regard then, as movie stars and rock stars are today. It was the task of those steamship masters to transport millions of board feet of lumber, many tons of iron ore, coal, and thousands of bushels of grain that was feeding Canadian and American families. Without those brave merchant sailors, the Great Lakes region would have likely remained a remote forested area.

Captain Johnston on duty / photo J. Brandow

The snow falling outside of Captain Johnston's library window was coming down in cotton ball size clumps, little

Alexandra watched the winter show from her seat at the kitchen table. As she made her way through her dish of peaches, Mother sang softly a few feet away as she continued to stir her pot of stew. For now, there were no boats to load, no crew to supervise and no shipboard orders to give. The Lakes were frozen shut from New York to Port Arthur, Cleveland to Chicago. With little more than 12 weeks at home a year with his wife and children, it would take the captain at least a week or so to adjust to his surroundings.

Captain Johnston's library was his refuge; it was one of the largest rooms in the house with its high ceilings and tall windows that allowed the brief winter sunlight to pour in during those short days. Johnston spent a great deal of his time at home behind his massive roll top desk, reading and reviewing the bookkeeping tended to by his wife during his time at sea. But like all men, hoping to keep their better halves happy, the captain would also spend time in the kitchen with his wife Nettie, sometimes helping with the cooking and sometimes as just an observer. Alex would rest on his elbows at the table and lean slightly forward holding his saucer with both hands as he sipped Nettie's fresh tea. It was a standard practice of the captain to pour his brew from the cup into the saucer for it to cool. Nettie would pour herself a cup of tea and sit at the table with her husband and the two would spend hours talking about the many days that had gone by while he was at sea, and she with the children.

Nettie loved and enjoyed her children, but had to give up her career as a teacher to begin her family. While it doesn't seem logical or reasonable in modern times, female schoolteachers of that time could not be both married and employed. In most cases it was simply understood that the man was the breadwinner of a proper family and the woman was to stay home and cook and clean for her husband and rear his children. But no matter what the proper etiquette of the day called for, Nettie chose to raise the captain's children, and command the house.

The couple would share many smiles over the stack of post cards that had been sent and received from ship to shore during the past year. The captain and his wife would also bristle with pride as the reunited couple reviewed the report cards of their children's progress at school. Nettie kept her wooden postcard box in the kitchen. When the boys were in school and her daughter was napping she would often re-read each card while dinner was cooking. The captain would bring home his box full of post cards sent by his wife while he was away at sea. One of the few cards that still exist shows the Johnston's youngest boy in a familiar turn of the century pose.

One of the few remaining post cards exchanged between
Captain Alexander Johnston and his wife Nettie /
photo J. Brandow

It's a birthday greeting from Nettie to her husband. The boy dressed in Indian garb is the couple's son Neal. "Firefly wishes you many happy birthdays," dated December 9, 1909. The card was mailed in care of N.W. Mills N. Tonawanda, N.Y. to the steamer *James Corrigan.*

This was a special time when Nettie and Alex could laugh and share the family events that had slipped by while Father was away on his ship. It was also a time when a proud mother and father could gloat with each other over the growth and progress of their children, Frank, Chuck, Neal and Sis.

Of course, when a man is away from home for a great deal of time, bad habits can creep up on a person. Nettie had her hands full rearing the children and teaching them to become self-sufficient, but husband Alex had some interesting traits to deal with as well. Nettie would often have to rein in the good captain for several shipboard habits that would certainly not be tolerated in Nettie's house, much less in her spotless kitchen.

For Alex is seems, like many men of the time, had a habit of chewing tobacco and with the practice comes the art of spitting. The captain would take a good cigar, bite off a piece and tuck it between his cheek and gums. Spitting, as any real man will tell you, is just part of the ritual of chewing tobacco. Nettie was careful not to leave the captain alone for too long a time when he first returned home from his boat. Nettie would allow her captain to "chew" but would not allow a cuspidor in her home. In a less civilized domicile, the appliance was called a spittoon. So, while seated in his favorite kitchen chair with a mouth full of chew, and with Nettie out of the room, the good captain would test his luck, and his aim for the firebox of the woodstove. Alex often times did not walk up to the stove and remove the cast iron cover from the top burner to rid himself of the tobacco soaked spittle. Instead, the captain, while seated in his chair at the kitchen table, would take aim from half way across the room and attempt to execute the proper trajectory required to find his mark. Unfortunately for Alex, and the children as well, Nettie had a hunch when things were just too quiet. "Alexander Johnston! I will have none of that spitting in my kitchen." It seems the captain had hit his mark, placing the soggy projectile in exactly the spot he was aiming for, while he was still seated at the table.

"You march right outside and take care of your business, Captain!!!" Nettie was a commanding figure at five-foot tall, weighing nearly one hundred pounds, giving orders to a man twice her size. The captain knew what was best for him and what he needed to do to keep the peace, and with an apologetic nod to his lovely wife, Alex would exit the back door and take care of business.

Many of us may remember the time, when children were seen and not heard, and did not question the authority of their parents. Such was the case in the Johnston home. But no matter how strict, no matter how formal, people are still people, and like most of us, we may tend to be a bit more informal when we are behind the closed doors of our own home.

The captain, being the ultimate authority thought he was both out of sight and out of earshot of his children. Little did he know the heat register above the kitchen stove, that allowed the heat to rise to the second floor, often would serve as a hidden gallery for a young, invisible audience. The captain's children would tell you that if you were quiet enough, and could lie very still on the floor near the heat register upstairs, all four children could eavesdrop on the activities and conversations going on in the kitchen.

Trying to keep from laughing out loud when Mother was scolding Father for spitting in the stove was a tough job. A child knew there was a whooping coming if they were to give away their position. After all, Father didn't know he had an audience on the second floor. And the children knew they would be in terrible trouble if Father knew they were secretly listening. The children all knew their best protection from being detected was to bring a pillow to laugh in, just in case something funny would happen down in the kitchen. At least you could cover your mouth and attempt to muffle the roar of laughter as you tried to quietly scurry away. The captain had dealt with the problem of nosey children in past years and made it a steadfast rule in the Johnston home that bare feet were not allowed. This rule made great sense when it came to preventing foot and toe injuries. The clomping sound of the children's shoed foot steps also served to alert concerned parents as to their children's exact whereabouts in the house, for whatever reason.

The Johnston family photo album serves as visual record of the captain's growing brood and an opportunity to track the styles of the period. In October of 1907, the Johnston family welcomed a new addition. This time it was a girl, and in those days babies were born at home. The Johnston's' new addition would be christened Alexandra, a namesake for her father.

Frank, Charles, Neal and Alexandra circa 1909 /
photo J. Brandow

But Alexandra, being very much a Johnston and born with a strong will, would begin to question why life had to be so complicated. It would stand to reason that the budding young lady would want to know "who that man in their house was," after all, the captain hadn't been home in a year, and there were a few years when he did not make it home at all.

Alexandra was only a few years old when she was faced with her first social dilemma. Alexandra was quite serious when she complained to her mother, that her name was too difficult for a small child to pronounce. From a child's point of view, it made sense, the name Alexandra had four syllables, and was quite a mouthful for a youngster. And, was it fair that her brothers had one syllable names that were easy to pronounce, while hers was somewhat of a chore?

While Sis and her mother were in the kitchen discussing the problem, Nettie came up with a solution. "Why don't we have your brothers call you sister? Would that be O.K. with you?" she said. Well, sister is much easier to handle with the tongue than Alexandra, she told her mother. But the boys had a say in it too, they decided it was much easier to call their little sister, "Sis," and it stuck. So, "Sis it shall be," said mother. Sis was now one of the guys.

As Sis recalls, her father's homecoming at the close of the shipping season was rather ceremonial by today's standards. "Upon Father's arrival, the boys would line up by the door at the front foyer and bow when Father entered the house saying *welcome home father*, girls didn't bow they curtsied. The man of the house was the master of the house, or at least that's what his wife Nettie allowed him to believe.

There were many rules for the children, and many lessons they were expected to master. While Father was away, the boys were to help with the household chores, and were to take lessons from Mother on sewing, embroidery, cleaning and cooking. When Father was home, the boys tended to more manly tasks such as shoveling coal into the furnace and removing the clinkers and ash from the spent coal. "Father says a man should be able to do anything a woman can, except have children." A lesson his sons, years later, would also employ in their homes, with their children. But as youngsters, the captain's children would also get a taste of rigid discipline from their father. How could a sea captain command men and a ship without some of that leadership spilling over into his home life? "The children knew there were rules and breaking them would have a heavy penalty," Sis reminds us. "Luckily for the children, Mother would be there to confer with the captain and remind him that he is not commanding the men on his ship, he is raising children."

But, when discipline was to be handed out, the children were instructed to find their own switch. Sis would head for the basement to find hers. After a thorough search, Sis would return with her chosen article of discipline. In one particular incident of lesson learning, Nettie, after receiving her daughter's selection from the woodpile in the basement commented that "she had never seen such a small piece of wood in her life." It was also the practice at the Johnston household that there would be nothing between the switch and the intended target of a child's behind so the message could be delivered directly. Right was right and wrong was wrong and it was the most important lesson a child needed to be taught. The main lesson at the captain's home was, it too was to be run as a tight ship.

Captain Johnston earned his "Master's" license with dedication and hard work, it would be this lesson he would impress upon his children. The last license issued to the captain has been preserved and is currently part of the collection of the Saginaw River Marine Historical Society.

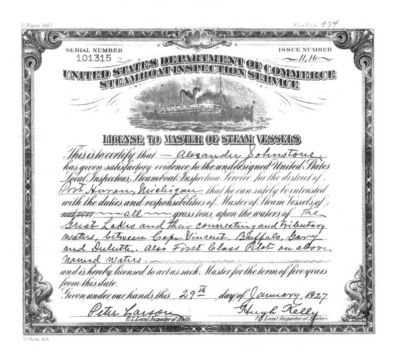

Johnston Shipmaster Certificate /
courtesy Saginaw River Marine Historical Society

Included in the Johnston file is his Lake Carrier Association certificate, renewed before each sailing season.

Captain Johnston had been disciplined as a sailor from a very young age. A short biography of Alexander Johnston appears in the "History of the Great Lakes" Volume II (page 490).

Captain Alex Johnston, a well-known and popular shipmaster, sailing out of West Bay City, formerly of Detroit, is so diffident about speaking of himself that it is difficult to render justice to his hospitable, upright, conscientious traits of character and disposition in an article of this nature.

He cannot however, preclude the writer, who has the pleasure of his acquaintance, from saying that he is a courteous, compassionable man, of fine physique, warm and sunny temperament, and generous to a fault.

He was born December 11th 1856, at Moore Ontario. a hamlet on the St. Clair River, and is of Scotch-Irish decent. He received a liberal education in public schools of his native place, and the studious habits acquired during his boyhood days are still retained, and it is unnecessary to say that he is an unusually well informed man relative to general history, as well as current events of the day. He had no artificial influence to help him obtain the position he now holds as master of a steamboat, but has secured it by his own merit.

Soon after reaching his majority Captain Johnston began his career on the lakes, his first boat being the Jupiter, on which he remained for a short time.

He then held various berths, from that of sailor before the mast to the office of mate, up to the spring of 1894, when he was appointed master of the steamer Isabella J. Boyce, trading between Bay City and Chicago, and from Buffalo to Lake Superior ports.

Isabella J. Boyce / photo Ralph Roberts

Captain Johnston in his quarters / photo J. Brandow

In the spring of 1895 he was appointed master of the steamer Sparta, engaged in trading to all ports on the lakes, which he sailed with good business success for four seasons including that of 1898.

Steamer Sparta / photo Ralph Roberts

Socially, the Captain is an ardent Mason of high degree, having reached the thirty-second, and is Noble of the Mystic Shrine. He is also a member of the Ship Masters Association. And carries Pennant No. 1081

In 1897, Capt. Alex Johnston was wedded to Miss Nettie S. Neal the eldest daughter of Capt. William Neal, of West Bay City. The family homestead is pleasantly situated on South Dean Street, West Bay City, Michigan.

The article on Captain Johnston was written when he was in his early 40's and still working his way up.

Alex Johnston was one of eleven children growing up in Canada. Crossing the river to the American side, Johnston became a U.S. citizen and like all good men of character joined the Masonic Lodge; as his documents read in 1894 while still a "mate" on a steamboat.

In the coming years Captain Johnston would transfer his membership to the Lodge in Bay City, Michigan. His membership records remain in the organizations files today.

Captain Alexander Johnston worked very hard to get to the top, to command his own ship. The sea was Captain Johnston's first love, the sea was also his life's adventure. Fortunately, many of the Captain's records and personal belongings are still in tact.

As a single man, Alexander Johnston began his sailing career by crossing the St. Clair River to work at the port of Detroit. The blossoming career of the sailor with so much promise would take him north to the port of Bay City, Michigan.

At the time, Bay City was the third largest city in the state, behind Detroit and Grand Rapids. It was also the only port between Detroit and the Straits of Mackinaw that was big enough to put a lake steamer in dry dock for repairs or refitting.

With so many sailors and lumberjacks who called Bay City home, it was no wonder the city, sliced in half by the Saginaw River, would also become one of the wildest cities in the country. There were plenty of jobs available no matter what skill a man had, or a woman for that matter, as well.

Alexander Johnston took up residence at the Campbell House, a huge three-story rooming house on the southeast corner of Third and Water streets. It was Water Street that beckoned both sailor and lumberjack with idle time, the lust for adventure, and a pocket full of money. A man's loneliness and sobriety could both be cured with a few dollars and even fewer steps. For this is the place they call "Hell's Half Mile." From Third Street south, Water Street was lined with saloons and bordellos with their crimson street lamps, guiding those seeking the company of the "devil's daughter."

This is also the reported intersection where the real Paul Bunyan is said to have met his maker. Bunyan, whose name in real life was Joseph Fournier, was reportedly murdered at the foot of the Third Street Bridge. The span of iron girders linked the two cities of West Bay City and Bay City together; the structure was said to be the third largest swing bridge in the world. Built by the Milwaukee Bridge and Iron Works in the late 1800's, the Third Street bridge would serve horse-drawn traffic, and later, automobiles, until 1976, when she would finally fatigue and fall to her death.

Third Street Bridge / courtesy Leon Katzinger collection

Retired newspaperman, historian and author, D. Lawrence Rogers has researched the legend and the facts surrounding the Paul Bunyan story, America's best-known lumberjack of lore and life. Rogers presents some interesting facts passed down over the years. The veteran author sheds new light on the life and the murder of the famous woodsman. While Bunyan's story has its roots on the banks of the Saginaw River, his life played out on rowdy Water Street, better know as "Hell's Half Mile."

In regards to Captain Johnston, who lived across the street from Paul Bunyan's death scene, there is little known about any alleged activities the good captain may or may not have participated in, during his stay at the Campbell House. The structure itself still stands today. Remnants of the letters that spell out *Campbell House* can still be seen beneath a thin layer of flaking paint, on the west side of the three story brick building. After serving as boarding house and later a furniture store, it is now one of the largest antique shops in the Midwest.

A chair from his apartment was in the possession of his daughter, until she passed it on to the writer of this project. Johnston's daughter says "The Captain's Chair" came from the place her father once stayed (Campbell House) before marrying her mother.

In late December of 1897, Alexander Johnston married Nettie Neal, a year after the death of his new wife's father.

The Bay City Tribune
December 30, 1897

Captain Alexander Johnston of the Steamer SPARTA was united in marriage last evening to Miss Nettie S. Neal, by Rev. T.W. MacLean of Trinity Episcopal Church.

Post card of Trinity Episcopal Church / photo Leon Katzinger

Trinity Episcopal Church remains, as it was. It's wooden trussed interior is reminiscent of medieval England. The

Johnston's marriage certificate had been preserved by the family and donated to the Saginaw River Marine Historical Society.

For a short time, the Johnston family lived in West Bay City. But this captain of status and station, with his growing family would need a proper home that would meet the needs of his blossoming family.

It was finally agreed, with prompting from his wife, that the family would relocate to the new home on Ninth Street. With plenty of room for their family, and plenty of yard for the children to play in, this would be the Johnstons' home for decades to come.

The library was a large room on the first floor facing the east; the morning sun would make it the perfect place to read with the morning's tea. It would also be the room where the children could study at the large round oak table. Across the room was Father's massive roll top desk, next to it, was the captain's favorite reading chair, the same chair the captain's daughter treasured, until her death. Also in the library were many bookshelves, protected from dust and the elements with glass doors. There were huge double pocket doors between the library and the sitting room with its oak mantled fireplace. Upstairs, was the master bedroom, and a room for each child. Needless to say, this was a modern house. Its gaslights gave way to electricity, it had indoor plumbing, with bathrooms on the first and second floors. "Quite an elaborately furnished home for the time," said Sis.

The Johnstons were doing quite well at the turn of the Century. The 1900's offered much promise for the future, and Alex's reputation for being a competent captain led to many promotions, to larger and larger ships, and bigger and bigger paychecks.

When steel vessels began to replace the aging fleet of wooden hulled steamers, Alexander Johnston was among the first of the captains to command the new breed of ships.

James Corrigan / photo courtesy Ralph Roberts

The *Corrigan* was nearly two hundred feet longer than the largest wooden hulled vessel still in operation on the lakes. After overseeing her sea trials and several years at her command, Johnston drew a new assignment.

The *Powell Stackhouse* was the new command of Captain Johnston. At a length of 504 feet, the *Stackhouse* was one of the larger Great Lakes carriers transporting goods to major industrial inland ports.

Powell Stackhouse / photo courtesy Ralph Roberts

With larger boats working the lakes, it was now more convenient for the captain to have guests aboard, namely visits from his family. During one of those shipboard visits, the captain's daughter recalls spending several weeks aboard the *Stackhouse* with her mother, aunt and her brothers.

Left to Right: Aunt Mae, Sis, Nettie, Neal,
Charles and Frank / photo J. Brandow

This photo was snapped by Alex Johnston behind the *Stackhouse's* pilothouse. Between Sis and Nettie you will notice the pendulum type device that will measure a ship's list. (The tilt of the ship)

Next it was time for another one of those dreaded photos. The children, especially Sis, hated the way they were always told to line up, from tallest to shortest, for their pictures. Sis said, she always felt like she stuck out like a sore thumb, always being the shortest of the Johnston crew. And being the youngest, there was no way she would ever catch up with the boys. She would always be last in line for the picture. While it was difficult for a child to behave all of the time, daddy's little girl often bent the rules. While running down the port side (left) deck of the *Powell Stackhouse*, Sis slipped and fell, cutting a big gash in her leg from the sharp corner of the steel hatch cover. Each and every time Sis told the story, she would hike up her pant leg and show off her scar, as if it were a badge of honor, or at least proof that she survived the ordeal.

Sis was brave, or at least she wanted her father and his entire crew to think so. Once the bleeding was stopped, and once her leg was patched up, things were back to normal. There were times however when the family could not be together. Sis recalls a time or two, when there was a merchant sailors' strike on the Great Lakes, and Father would have to stay onboard his boat, while watching the owner's fleet tied up at the docks, for fear that disgruntled hooligans would cause harm to the vessels as they sat idle and lifeless.

Captain Johnston, despite any Victorian quirks he had at home, truly was a captain's captain. He lost no vessels, he had no collisions, and he was never cited for violating maritime rules or regulations. And he always turned a handsome profit for his employers.

Chapter Three

Nettie is the Captain of this ship!

With the new age, came many wonderful things. The Johnstons were among a handful of families who had a telephone for regular service. The new invention of Edison called the phonograph also became an added feature of comfort and status at the captain's home. "But the phonograph would never take the place of Mother and her violin," said Sis. "Once the children were in their beds upstairs for the evening, Nettie would tune her violin and play a lullaby for the children while standing in the front foyer. Sis said, "Mother's music would fill the house and make way for the sandman to softly send the sleepy children off to dreamland."

The Johnston Children / photo J. Brandow

Sis says, "When Father was away on his boat, and even when he wasn't, Nettie ran the show. It was her job to teach her children to cook, sew, embroider, dust and clean and all of the other tasks it took to keep a house in order."

But boys will be boys, and boys like to have a secret hiding place for their treasured things. Sometimes those secret places, were so secret, they were forgotten. Some treasure lie undisturbed for decades. A few of those toys were discovered more than seventy years after the family had moved out. In one case, under an attic floorboard, the handmade toy of a prankster.

The star-wheeled contraption was designed to make a racket, when scraped down the outside of an unsuspecting neighbors window. Sis said, "Mother would jump out of her chair when Chuck would play a trick on her. That thing made such an awful sound."

If a boy was quick and could run fast enough to get off the neighbor's porch, the prank was considered a success." But a prankster, who thought the jig was up, may try to hide such a piece of evidence, under the floorboard of an attic perhaps, where no one surely would think to look. Chuck, the Johnston's second oldest son was correct, no one did look, and his noise-making toy would remain out of sight until long after his death. It was one of those discoveries that a fix it up homeowner would consider a treasure to come across during renovations.

It was Sis who not only identified the handmade, starwheeled noisemaker, but also pointed the finger at her brother Chuck as its rightful owner.

Sis would also lie claim to a gold velvet wrapped horseshoe and a children's block as her own. The velvet wrapped horseshoe was from a school play. The children's block was from a set she had as a child. She believes one of her brothers may have hidden the items from her while playing a game of hide and seek.

Sis with China Doll / photo J. Brandow

"The boys always played hide and seek in the attic, even though Mother ordered them not to. The stairs leading to the attic and the roof hatch were far too steep for a young child to maneuver. They would surely get hurt or worse yet fall down that long flight of stairs that led to the forbidden hideaway," said Sis.

But, the allure of that hideaway, and its many hiding places was the perfect place for an older brother to play tricks on his younger sister. Especially when it involved a little sister's treasured doll she received as a gift from her mother.

Upon one of my first meetings with Sis, when I began researching the Johnston home, Sis had asked if I found her china doll. Puzzled at first, by her question that seemed to come out of left field, I was left wondering if she, in her advanced years had confused me with one of her brothers or even her father.

Quite the contrary, Sis knew that as the newest owner of her former family home that I would be poking around in every nook and cranny. And after more than 90 years, she still had not found or forgotten, the doll her brother Chuck had hidden on her while playing in the attic against her mother's will.

I told Sis I would keep an eye out for it. The doll in the photo that Sis is holding is thee doll that still remains at large. And as promised, "I will keep a sharp eye out for its eventual recovery and return, even after more than 90 years on the lam."

Before Nettie Neal's career as a captain's wife and a mother, she was a schoolteacher in Bay City, Michigan. Nettie loved children and taught the elementary grades.

She loved her job, and inspired one of her younger twin sisters, Lilly Mae and Violet Rose, to also become a teacher. It was Mae who answered the calling of the classroom. Violet had a different vocation; she became a much in demand dressmaker and created the best, for the best in town, and her sisters. Aunt Mae taught the elementary grades for many years. Aunt Mae and Sis had a special shine for each other; perhaps Sis would one day follow in her mother and aunt's footsteps, and become a schoolteacher too.

Only weeks after the tragic sinking of the *Titanic*, in April of 1912, life was beginning to return to normal. Winter was finally over and it was time to celebrate the longer days and warmer weather.

It was Easter and a young Sis does a little jig for her mother snapping the picture. In the background, wearing the big hat is Aunt Mae, calling to Sis to come inside Wenona School for an Easter presentation.

Sis at Easter / photo J. Brandow

"We had a very special relationship, Mother and I, we were friends," Sis says. So good a friend that Nettie would let her daughter know when she did not live up to her expectations. Sis was a Johnston, and the Johnstons followed the rules. Father's number one rule, was *following the rules*. But Nettie was also a Neal before marriage, and the Neals often bent the rules, so there was some room for flexibility at times, especially with Nettie in charge of the home ship.

One day, when Sis was playing baseball across the street with the boys, Nettie shouted out to her daughter, promising a silver dollar if she could hit a home run. The next pitch was a homer, right off the bat of Nettie's little girl. Nettie was filled with pride as the ball soared high into the air. Upon its return to earth orbit, the home run ball narrowly missed Mrs. Johnston, who was standing on the front porch. The homerun ball smashed through the front window, into the sitting room leaving glass shards everywhere.

Nettie told her home run hitting little darling that, "replacing that piece of glass would run, about a dollar." It would be a story the two of them would laugh about at family gatherings for years to come. Nettie, being the retired schoolteacher had a special way with children. "It was the Johnston house where all the neighborhood kids would congregate. The children all thought Mother was kind and gracious, she was, but the truth be told, Nettie also thought it was the best way to keep a close eye on the kids," Sis said.

Nettie knew how to give each one of her children that special attention that a child needs. The Johnston kids all loved the heel of a fresh baked loaf of bread. It was simple arithmetic, a loaf of bread had only two heels, and there were four children in her care. "Every Friday, Mother would bake a round loaf of bread and cut it in such a fashion that everyone got a heel and a pat of fresh butter to go with it," recalls Sis.

"Mother's shopping routine was quite different from what we are accustomed to in these days of Monster Supermarkets and mega department stores. In those days, walking was the prime mode of transportation. There were two neighborhood grocery stores near the house. These small family owned markets would offer the regular staples of sugar and salt and potatoes. There were canned peas and canned Pet milk for baking. Meat for the family dinner was sold a few blocks away in the Columbus Avenue business district."

Columbus Avenue Business District /
photo courtesy Tuthill family

The streetcar ran up and down Columbus Avenue, all day and half the night. If you examine the extreme right edge of the above photograph, you can catch a glimpse of the working streetcar. Everything a person would ever need or want, was right there on Columbus, the grocery store, meat market, hardware store, the apothecary shop, the flower shop, the blacksmith shop, and of course a watering hole or two. There were two meat markets Nettie would use to supply her meals." Her favorite meat market was the closest to the house;" Sis fondly recalls. Today, that same building is now home to Grampa Tony's Italian Restaurant. The eatery is owned and operated by the Lagalo's, the children of an Italian immigrants' child. The building still retains some of its early exterior demeanor.

"The other market, for fresh steaks, chops, and lard, was a few steps further, at the corner of Lincoln and Columbus Avenues, known at that time as Zeigler's," according to memory of Sis Johnston. The building is still in use as a neighborhood market today. It too holds on to some of the more familiar facets of the buildings original design. The store sign in front, now reads, "Tuthill Brother's Market."

Zeigler's Store / photo courtesy Tuthill family, circa 1900

While they no longer cut fresh meat, hanging on the wall over the counter is a photograph of "Old Man Ziegler" himself standing in front of his store with a horse drawn delivery wagon ready to go to work.

"Just imagine what it was like to haul groceries home from several blocks away with a fully loaded bag in each arm," Sis said. "Many people had hand carts, they could load to the hilt, and pull behind them as they walked home. That was progress!"

Like many of us, Sis learned some of her best remembered lessons the hard way. In this case, it was a child's first lesson in getting caught, sticking their fingers where they don't belong. Nettie would keep her spare change in a cup, tucked away on a shelf, beyond the reach of little hands. Nettie used the change in her cup to pay the iceman, the paperboy, and the man on the horse drawn fruit and vegetable wagon. Daughter Sis, who by this time is becoming more fully acquainted with her immediate surroundings, took note of Mother's cup, filled with shiny coins. While accompanying her mother to market, Sis noticed the practice of buying and selling goods with the use of money. "Mother would take money to the store and get the things that we needed for dinner," Sis said.

With a cupful of money on the closet shelf, and Sis now big enough to reach it, she would now be able to employ the business of economics, the law of physics, a dash of imagination, and a small dose of effort to pull a dining room chair over to the shelf. "It would be an exciting adventure indeed, a one time adventure led by a sweet tooth. Pulling that chair to the cupboard, and reaching inside for just a few coins, Sis knew she could buy enough candy for herself and all of her friends, to last an entire afternoon."

The storekeeper sold a dollars worth of just about every kind of candy the store had to offer. There was licorice, jawbreakers, chewing gum and peppermint sticks. Life was good as Sis enjoyed the spoils of her newfound windfall with her friends. But it would be technology that would be the downfall of Sis's well-executed excursion. Following the sale, the storekeeper called Nettie's four-digit phone number and asked if "Sis" had received some money for a birthday or other occasion. Puzzled by the inquiry, Nettie craned her neck and looked out the window to see her daughter's slow return home. "Mother thanked the storekeeper for keeping an eye on her daughter and hung up the phone," Sis recalls her mother saying years later. Little did Sis know, the celebration of sweets would not last, once she got home.

For Sis, the moment of truth had finally come. "Where did you get a dollar?" was the question Sis was met with at the door. "Why, I got it from that cup full of money in the cabinet", pointing to Nettie's hiding place. "That's my money! Did you ask? Who said you could take it?" Sis recalls, "Mother had me do chores for a long time to pay back all of the money I took from her cup, I dusted and cleaned the house for a long time."

Too cold to be outdoors, Sis and her brother Neal were playing in the sitting room. They had to be careful, not to knock anything over, while they were playing.

"Mother was in the back of the house in the kitchen preparing dinner. Growing bored, Neal and I decided that flying through the air was the best idea they could come up with at the time."

Jumping off the arm of a chair to the floor below, gave the youngsters some wings, but they soon tired of that and moved on to yet another method of flight.

"What would happen if we used Mother's big rocking chair, to jump from? Neal could rock the chair back and forth and when the rocker was nearly ready to flip over backwards, I would jump forward off the chair and get high in the air we were having a ball," Sis recalls. The theory was put into practice and the chair rocked and rocked and the kids flew, higher than ever.

The trouble was, standing on a rocking chair, that your brother is slamming back and forth is difficult at best. When it was Sis's turn again to fly from the seat of the rocking chair, she fell over the side, smashing onto the floor in a most awkward position and broke her arm.

Mother came to her aid; she could deal with the children's disobedience later, she needed to take care of her injured daughter right now. After assessing the injury, Nettie called Doctor Tupper's office a few blocks away on Center Avenue, near Trinity Episcopal Church where she got married.

Doctor Tupper was a popular and very kind man and very good with children. "Mother said, I was going to have to be brave and we would walk down to the Doctor's office. But first she made a very strong cup of tea for me to drink before we left."

With the medicinal practices of the day, Sis believed she had been served a panacea, easing the pain of her broken arm, for the walk to Dr. Tupper's office.

Rubbing her arm in the place it had been broken more than ninety years prior, Sis said, "He looked at it, and before I knew it had set my arm and covered it with a plaster cast. I got some candy, and a ride home in his new Cadillac." Sis said she, "asked the doctor to take the long way home."

Chapter Four

"Sis" The Captain's daughter

The rising plume of thick smoke that was billowing from beneath the back porch sent an alarming chill up the spine of Captain Alexander Johnston as his taxi pulled to the curb, delivering him home for a few day's vacation. The cabbie would have to wait a moment to collect his fare as the captain wasted no time in jumping out of the motorcar to begin investigating the source of the apparent fire at the back of his house. This was certainly no way to begin a few days rest at home while his boat was in dry dock for refitting and repairs.

The captain ran up the drive, to the back of the house, and stooped to locate the fire's origin. What he discovered were a pair of tiny eyes looking back at him through the smoke, as he peered beneath the wooden porch. The entire neighborhood could hear the captain say, "What the hell are you doing under there?" Breathing a shallow sigh of relief, Captain Johnston thanked the Lord that his house was not on fire. But he was extremely unhappy with the discovery of his 5-year-old daughter playing with a box of match sticks. The captain paid his cab fare and set his bags down in the drive to deal with his adventurous daughter, Sis.

"Father never swore at me before, so I knew I was in big trouble and my heart felt like it was going to pound right out of my chest," said the captain's darling little girl. Again, the captain asked the question, "What the hell are you doing under there?"

"Father caught me smoking leaves under the porch, it was my first time! I saw the boys doing it and it looked like fun, so I thought I would give it a try," Sis said. The captain however, was not the least bit amused. "Show me how you did that," barked the captain. "I told Father it was easy, I took some toilet paper and wrapped up some leaves and lit it up, I saw the boys doing it too! Father stood there shaking his head, he said some words I didn't hear very well and couldn't understand. Father said he would have a long talk with the boys and told me never to do that again, and I didn't."

"Oh, I was so scared when I saw his face looking at me under the porch, I didn't know he was going to be home, and I knew I was in for a good licking." Ninety years after the fact Sis Johnston recalled the event as if it were only yesterday. "You don't forget things like that, especially when it's Father that catches you at it."

Alexandra "Sis" Johnston shares some of her treasured memories and some of the not so treasured lessons she learned the hard way. "Sis" Johnston would eventually reach out and touch many thousands of young lives in her career as a teacher at Bay City Central High School. Miss Johnston was known as a tough, no-nonsense type of person with a good sense of humor, who also conducted her classes in rigid old-fashioned style while keeping her students on a short leash. Miss Johnston grew up in a household where rules were rules, and discipline applied if the rules were violated. School was about learning, and it was her job to make sure her students did the very best they could. After all, Miss Johnston was taught by the best, her mother and her Aunt Mae were schoolteachers. Sis Johnston only hoped to do, "as well as Mother and Aunt Mae when it came to classroom instruction, and the success of her students."

Miss Johnston would not only employ her skills as a teacher, she would also employ many of the hard taught lessons she learned, following the rules of a strict sea captain. Sis Johnston's insistence on student excellence was an effort to get her students to expect more out of themselves, and not be satisfied with anything less then the very best they could do. Miss Johnston would also commend and acknowledge the success of her students for a job well done.

Sis Johnston was born in October of 1907 at the family home, just one year before Henry Ford introduced his first Model-T automobile to the public. Sis would be born into an amazing world of monumental change; she would bear witness to the Great Depression, two World Wars, the invention of Radio and Television and the unthinkable feat of a man walking on the moon and the Internet. The developments and inventions of the time were nothing short of miracles. While some people of her day did not welcome change, like many in our time, Sis welcomed new things and new ideas.

The invention of the flying machine, dreamed of for centuries by scientists and inventors, began unfolding during her lifetime. The first sustained powered flight of a human was accomplished by the Wright Brothers only 4 years before her birth. We too are witnesses to history unfolding as we witness the flying machine still being revised and improved upon today.

Admiral Perry would reach the North Pole in 1909, an event thought impossible for the time. Thomas Edison would show off his new invention of "talking pictures" in 1910, but it would still be another 17 years before "talkies" would take over silent films. Sis was 5 years old when the infamous *Titanic* struck an iceberg off the coast of Newfoundland, in 1912. A sailor was a sailor and the ripple effect of such a tragic event would touch the souls and break the hearts of all who sailed, and their families.

Worrying was for grown-ups and in the early days of the 1900's there was still time for a child to be a child, once their chores were done, of course. Sis Johnston would always be the little sister to her three older brothers, Frank, Chuck and Neal. Never wanting to be left behind or left out of the boys more adventurous activities, Sis did what she could to keep up. Sometimes the wrath of her exploits far outweighed the adventures of her brothers. It was one of the boys who taught their little sister how to strike a fire using a match stick. Perhaps the little girl had only to watch the example set by her older brothers, who made smoking leaves and lighting matches look like a lot of fun. "For a child, there is nothing more amusing or captivating than striking a match for the very first time," said a smiling but remorseful Sis Johnston.

Always the precocious lass, Sis set off to explore the world of being a grown up, like her brothers, like her mother, and like her Uncle Walter. The Johnston children knew it was less than wise to let Father or Mother, catch them breaking the rules, they knew they would have to pay. In the heavily starched Victorian era, Sis says, "it was more often than not a naked rear end would meet the business end of the switch. But that's how things were then, tough rules and tough punishments."

Being a kid was not always easy, during Victorian times. As Sis recalls, "children were seen and not heard, and talking back to either one of my parents was unthinkable." Many homes had a designated razor strap hung on a nail in the pantry, a gentle reminder to the children that "The Strap" is not too far out of reach." At the Johnston home, discipline was a little more creative. "It was a time when parents spoke only once, and the children not only listened the first time, they were motivated, with good reason. But even after a lesson is learned, you still had to walk softly for a while," Sis remembers.

Nettie was busy in the kitchen, preparing dinner, when daughter Sis, after her leaf-smoking episode, comes in to ask her mother for a matchbox. "Not a full one Mother, an empty one, please?" "Young lady, what do you want a match box for? Didn't you learn your lesson the last time?" "Mother, I want to use the empty match box to bury my goldfish in. It's dead and I don't want to flush it down the toilet, OK?" "With a sigh of relief, Mother emptied a matchbox for me to use for the funeral of my goldfish."

Sis says, "it was a simple ceremony that took place on the west side of the house, under the kitchen window." She carefully prepared the makeshift coffin and placed the goldfish, just so, before she slid the lid closed. Sis's closest friends attended the simple but solemn ceremony as she laid her goldfish into eternal rest. "The proceedings of course were conducted under the distant but watchful eye of mother, who was just making sure, that ceremonial leaf smoking wasn't a part of the goldfish's funeral ceremony," Sis said with a smirk.

Trying to emulate her older brothers, Sis would land in hot water from time to time, but those tomboy notions would soon give way to girl stuff and other lady like things. Sis's biggest worry, at one time, was getting her buttonhook to work on her shoes. "Mother had the touch and could button up her shoes faster than anyone I know!" Keeping up with styles was Mother's concern. She was always a modern woman, and fashion had always been her passion." At the time, Sis says, "she didn't worry about things like fashion; she wanted action packed adventure like the boys had." Often times she would push aside her little girls' books and crack open the coveted action novels of her brothers. The adventures of "The Motor Boys" and "The Rover Boys" were more to Sis's liking; "They had more action! And I liked that!," proclaimed the aged adventurer.

Sis loved action; she would instantly become the "Dare Devil" of the block, once her roller skates were clamped onto her shoes. While rather crude and antiquated when compared to today's composite inline "roller skates," real roller-skates, were made of pressed steel. The contraptions had to be clamped to the edge of the shoe's sole and tightened down. You were a rotten egg, if you weren't one of "the kids" with a roller skate key tied on a shoelace hanging around your neck. For the time, in children's circles, Sis recalls, just how important it was to wear her roller skate key, on a long shoelace. "Some skate keys were plain and utilitarian, others, the more preferred, with added features, could perform a number of tasks. You could not only lengthen, or widen the skates, with a key, it could also be used to fix or replace the steel wheels. If the leather ankle straps were too tight and the buckle was difficult to release, the best skate keys had a hook to help undo the buckle latch."

One of Sis's best skating partners was her friend and classmate, Margaret Large, who lived dutch corner (katty corner) from the Johnston house. The girls would spend hours roller skating, jumping rope and drawing chalk lines on the sidewalk to play hopscotch. But Sis reminds us that things in her early days were much different than they are today. "It was most unusual for a child to have money in their pocket. Kids had to find things to occupy themselves, with something that did not cost money." For Sis and Margaret, playing paper dolls was one of their favorite activities. "In those days, we used our imagination! And we had a lot of fun too!"

For those who never had the pleasure, paper dolls, I'm told, usually consisted of a girl or a young woman made from a colorful cardboard cut out. The doll had a paper wardrobe, complete with tabs to bend around the edges, to keep the paper clothes in place. This "paper" coat was discovered wedged between a wooden baseboard and the wall during renovation. Sis's eyes were as big as saucers when she laid eyes on it, re-discovered, more than ninety years after it was lost.

Paper doll coat / photo Glen Groeschen

By the age of 10 years, Sis was becoming aware of world events. The United States entered the First World War in

1917; the Russian Czar had his best days behind him as the revolt was underway.

In 1918, "the war to end all wars" was finally over, but death was still headline news, the world flu epidemic, spread more than just fear, it would eventually kill 22 million people. The next year would bear witness to the "Black Sox" baseball scandal, where several players admitted they were paid off to throw the World Series, and growing corporate giant General Electric would form a new company called RCA.

The "Roaring 20's" would usher in the modern age and with it, turn America upside down. Prohibition became law, driving those with a thirst for something stronger than soda pop underground, to the speakeasy. Many men of the era were throwing their arms in the air with disgust as women voted for the first time in national elections. And that new fangled contraption called Radio, or Wireless as it was better known, was making its way into American homes at an alarming rate, at the cost of ten dollars a set.

The 1920's were a Godsend to women, or at least it was their first opportunity to begin spreading their wings on numerous personal and political issues. Joan of Arc was canonized as a Saint and the first "Air Mail" was delivered from New York to San Francisco.

About the time Egyptian King Tut's tomb had been discovered and opened, in the early 1920's, Sis had become a young lady. One of the many things I came to admire about Sis was her knack for organization. Here in her eighth grade class photograph, Sis had the foresight to write down the names of each one of her school chums.

While it's probably a safe bet to assume most of Sis's classmates have since passed on, the writer wishes to share the names and the faces captured on film, in an effort to perhaps aid their descendants and the genealogists in putting a face on history or their personal heritage, and for the future generations who have not yet joined us.

left to right, beginning with the back row. Frank Smith, Ray Strange, James Cuthbert, Robert Gouse, Henry Durbe, Raymond Beird. Second Row (from back): Ada Madgesein, Ruthby Cook, Helen (Toyzan)?, Francis Shultz, Alexandra (Sis) Johnston, Opal Smith, Dot Taylor, Howard Cook. Third Row (from back): (Talo?) Neal, Nellie Yeamans, Betty Thomas, Francis Rowely, Elanor Timm, Irene Mason, June Cuthbert Front Row: Robin Erwin, Francis Tempkins, Hubert Ames and Hariet Weber.

"The Johnstons were athletic minded people; Alexander was a very strong man, fit and trim well into his later years." says daughter Sis. The boys too were active, except for the baby boy, Neal. The youngest of the three Neal had been a sickly lad since he was born; he would later die, at an early age from tuberculosis. "Perhaps it was Neal's condition that prompted my brothers to strive for athletic perfection," said a pondering Sis Johnston.

In her high school years, Sis took part in extra-curricular activities and was involved with sports playing tennis and basketball. But, in those days, the girls had many obstacles to overcome in the sporting world. Remember that this was almost eighty years before the unthinkable would happen when females were allowed to suit up and participate side by side on the boys' athletic teams.

According to athletic instructor Sis Johnston, "high school girls were allowed to play ball in the tenth and eleventh grades, but in their senior year, it was not allowed. It was thought at the time that real ladies do not jump around on a hardwood floor. It was also thought by many school administrations that sports would be far too taxing on the bodies of the nation's future homemakers, as sports may cause irreparable harm to the reproductive systems of those who were expected to get married and raise a family, once out of school."

In Sis's sophomore year (1924) at Bay City Central High School, she played ball, and "was proud to do it".

left to right beginning with the front row: Isabelle (Turotte), Maria Taylor, Wanda Zempler. Second Row: Mary Martin, Gladys Beebe, Laura Rosenbury, Harriet Dehn, Adeline Erbel, Lucy Shaw, Sis Johnston. Third Row: Amanda Anderson, Kay Meyer, Betty Thomas, Irene Rutzen, Dorthy (Rubetfor), Charlotte Mitchell. (blouses and ties circa 1924)

Disappointed, she was only being allowed to play ball just one more year, Sis began to direct her energy in the direction of college. Sis, strong willed and determined, (a Johnston family trait) decided she was going to follow in the footsteps of her mother and her aunt, Mae Neal, and become a schoolteacher. This was Sis's last year in sports and she was going to make the very best of it.

Being an adult also meant Sis would be included in more adult events at home too. The captain and his wife loved to play cards, and often had friends over to play "500." "That's all they did sometimes was play cards. I hated it, playing cards was a waste of time." But Sis would, willingly, but reluctantly sit in, at the card table, if one of Mr. and Mrs. Johnston's card playing partners could not make it for the evening's event. But, often times during those long nights at the card table, Sis says, she would often find herself daydreaming about the things she really liked to do, "like taking pictures."

Sis's love for photography was fueled by the enthusiasm of her very good friend Gertrude Butterfield. Gert's father was a professional photographer and on occasion the girls would help out at the Center Avenue Studios of Alexander/ Butterfield Photography. In the days of black and white only photography, there was a method for adding color to the photographs, for a premium of course. It was Gert and Sis that would take the paint kit, and with the gentle stroke of the brush, they added life and color to photography. Sometimes the girls could earn a little extra money at the studio by helping set up the subjects with makeup, hairdo's and poses for their sittings. "The extra help would allow Gert's father the time he needed to concentrate on his camera and his work," Sis recalls.

Sis sat for her High School graduation picture at Gert's father's studio, revered to be one of the best, and most exclusive in the area. Genealogists, historians and other researchers advise that, photographs can offer much more information than meets the naked eye at first glance. With the assistance of computer enhancement, it appears that Sis's team photograph, was taken after school. The watch on the young lady's wrist in the first row indicates the photo was taken at approximately 3:15 P.M. While the time isn't necessarily significant, the discovery can serve as a reminder, to re-examine those family photographs, for the slightest of details, that may open the door of discovery.

From left to right, beginning with the first row: Florence Eichorn Second Row: Mary Logden, (Issy Turcoll), Maria Taylor, Mary Canfield, and Ima Collier. Back Row: Sis Johnston, Amanda Anderson, Coach Lounie, Irene Rutzen, and Laura Rosenbury. Circa 1925/Johnston Family Album

As we study these old photographs, it may help us to remember the world had become a more complicated place for a young person, as national and world events unfolded at an alarming pace. The message told by the young lady's wrist watch, is not so much what time it was, rather that this team photograph was not taken before a game. The photographs suggests that the "girls" posed for their team photo right after school was dismissed for the day. Sis Johnston said, "Florence was smart enough to know that you don't wear your wrist watch, when you are about to go out on the floor and play. I always told my girls to take off their rings and watches before they suit up."

No matter if it were next door or half way around the world, developing technology and global politics would touch their lives in some way. In 1924, Russian leader Lenin dies paving the way for Joseph Stalin to take power. Also in 1924, J. Edgar Hoover becomes director of the FBI. And a schoolteacher by the name of Scopes was convicted in the "Monkey Trial" for instructing students on the theory of evolution, instead of creationism. And sadly in 1926, it seemed the world had come to an end, at least for some, as silent film heartthrob, Rudolph Valentino was forever silenced, leaving his millions of imaginary lovers, stunned and heartbroken, in the wake of his passing.

With each year that ticked by, women were proving that they could do more than cook and have babies, much more. Women could do the same things a man could do, even better. Women of the world rallied behind Gertrude Ederle, the first woman to swim the English Channel. The event unsettled many traditional men, while at the same time giving the women a patron saint and the icon they needed to further the cause of women's equality.

Wanting very much to be a part of the world's many changes, its technology, and its rewards, Sis Johnston, along with her best friend and sidekick Gertrude Butterfield decided they would battle the world and its challenges together, through thick or thin; they would remain the very best of friends.

With Mr. Butterfield's passing, it was Gert and Sis that would take care of Butterfield's last photographic clients and develop his final photos. It was a business, and the girls were going to keep the promise made by Mr. Butterfield to have the customer's photographs ready on time, despite the death of the shop owner. Of course, this would be one of the adventures the two would take on together. "There was a tall stack of negatives that needed to be developed; they would be soaked in a series of several different solutions to make the images come out on the negatives." Sis says, "We didn't always know how things would turn out, sometimes they did, and sometimes we had to start over."

Sis had a certain knack for adding the color to those black and white photographs, she said she "enjoyed adding a little life and color, to the photographs that would certainly be around a lot longer than any of us." Thinking of the future by preserving the past, Sis Johnston donated her paint kit to the Bay County Historical Society, where it remains a part of its permanent collection.

The "girls" enjoy a moment in front of Mr. Butterfield's lens for a test shot in the studio. Serious bicycle collectors will take note of the National Bicycles used on the set. Sis said, "it was not unusual for the two of us to ride twenty or thirty miles in a day, it was good exercise for us."

Bay City, Michigan, was home to the National Bicycle Company, which sold its (highly collectible) premium cycles around the world. Some collectors claim, National Bicycles were some of the best ever built. The tire rims and handgrips were crafted from wood.

Gert and Sis pose for test photo, with their
"National Bicycles" / photo Anna Mae Wires

Following in her mother's footsteps as a modern woman, Sis Johnston began her serious hobby, becoming a "shutterbug," as a high school student in 1924. Sis practiced her newfound art when she could and when money would allow. Sis says she "began to see her world through the lens of a simple box camera."

Fortunately, many of her snapshots are still intact and with them we can look back at "the times" as she recorded on black and white film.

Sis's picture of the front of her house not only reveals the brick sidewalk and the landscaping style of the day, it also brings to light the political leanings of the Johnston household. If you take note of the lower left, front window you can see a campaign sign with the oval busted photos of U.S. Presidential hopeful Calvin Coolidge. Sis's photographs are not only precious moments frozen in time, they are tiny windows to the past.

Johnston home as it appeared in 1924,
note brick sidewalks and political poster
in front lower left window / photo Sis Johnston

Sis's hobby not only illustrates, but underscores, the reason to re-examine old photos for clues, that are often hidden in plain sight. It appears that Sis's interest in photography was more than the practice of a developing art. Photography for Sis was also the means to preserve many of life's most precious moments, sometimes life's simplest. Sis's mother Nettie was ill, and while everyone tried to convince themselves that Nettie would get better one day, "we all knew she was dying. Things like chemotherapy and open heart surgery were never dreamed of when Mother was ill", Sis regrettably uttered with a far away gaze.

The shrubs surrounding the captain's home come to full bloom each year during the last week in May. It seems rather sad and ironic that while the blossoming shrubs that Nettie had planted are in full bloom, while Nettie's cancer wracked body is beginning to wilt, making her appear much older than her 58 years.

Nettie Johnston, May 1924 / photo Sis Johnston

Nettie attempts to appear normal, as she is photographed on the side of her home. This last known photograph, of the prettiest girl in town, gives proof to the advanced stage of her disease. Nettie's sister, Mae, would also succumb to the same illness, a few years later in the early 1930's.

Responsibility is a heavy burden, and for Sis, there was plenty of hard work ahead, but she didn't complain. As Nettie grew weaker, Sis did what she could to help out around the house, whether it was cooking dinner or making repairs to the screen door off the kitchen. "Brother Neal was now in the advanced stages of TB with little hope for any significant improvement. Father was now in his early 70's and retired from the fleet. Neal, who got away with nearly everything from soup to nuts, in his better days, has the appearance that he is much older than his years." Sis knew that life was about to change, it had to, her parents were getting older, times were tough, and she was prepared to do what she had to do, when the time came to do it.

Nettie Johnston's medical problems were getting worse, much worse, prompting her admittance to the hospital in Ann Arbor, Michigan. "The doctors had performed what surgery they could, but had also informed the family that her time was coming to an end. There was little anyone could do, but say their prayers, and hope for a miracle. The doctors did what they could to relieve Nettie's pain and limit her suffering, "Sis recalled with a tear in her eye.

Under the influence of the opiates that were employed to dull her gnawing pain, Nettie wrote her last letter, it was to her daughter, Sis. "The letter was never sent by mail, it was hand delivered to me by Aunt Anne and Aunt Maggie, who had just returned home, from a visit with Mother, in Ann Arbor."

Some of the handwritten words are difficult to decipher, because of her medication, but the message is still very loud and clear.

Nettie Johnston's last letter, to her daughter, Sis.

*"Your two nice letters received and I would answer faster
but I simply can't. My hearing is failing more than when I
am at home. And the noise around here would drive us (our)
mad.*

*If all goes well I depart to go home tomorrow. If any one
wants me badly enough to come for me. Love to father and
Neal, Mother.*

Nettie Johnston's last written letter / photo Tim Brandow

Just a few days after writing her last letter, Nettie boarded
the train in Ann Arbor with her husband and sisters and headed

for her home in Bay City. "Mother wanted to spend her last days in her home, where she raised her family, gave birth to her daughter and spent some of the best days of her life." Never giving up hope, but falling victim to her disease, Nettie Johnston passed away, with those who loved her at her side.

The Bay City Times
September 23, 1928

Mrs. A. Johnston called by death/ Well Known Resident Formerly Taught in Riegle School.

> *Mrs. Nettie Neal Johnston wife of Captain Alex Johnston died at her home, 1009 Ninth Street, Sunday after an illness of five months. Mrs. Johnston was well known throughout the city, having been a teacher in the Riegle School previous to her marriage. Mrs. Johnston was born in Chatum, Ontario, and had been a resident of Bay City for many years.*
>
> *Surviving Mrs. Johnston, are her husband, well-known for many years as a sea captain, three sons, Frank of New York City, Charles, of Washington D.C., Neal at home and one daughter, Miss Alexandra, One brother, Captain Walter Neal of Bay City and three sisters, Miss Anna Neal, Miss Violet Neal, and Miss May Neal.*
>
> *Funeral services will be held Wednesday afternoon at 2:30 at the home, with Rev. Harold McCausland, rector of Trinity Episcopal church, officiating. Burial will be in the Elm Lawn Cemetery.*

The times were especially heartbreaking for the Johnstons. Alex retired from his commission, as a merchant seaman, in 1927, to be with his dying wife. Nettie was too sick to enjoy her husband's company, to sip tea at the kitchen table, or review old post cards. Nettie was in no shape to hold a hand of cards like she and Alex did for so many years together. The best the captain could do, was read the newspaper to Nettie, when she was up to it.

One of the biggest news stories of the time was the impossible feat of flying a single engine aircraft, solo across the Atlantic Ocean. As the dreams of Charles Lindberg came true, it inspired many others to also pursue their dreams. "If a man can fly across the Atlantic Ocean in a little more than a day, why can't they come up with a cure for Nettie's cancer?" Sis recalls her father asking. A fair question, but with an answer that was as far out of reach as the moon. It was far too late to second-guess the lives that had passed before them.

Alex, like his father-in-law and his brother-in-law was, by trade, a sea captain. He loved his work, he was better than the rest at it and it afforded his family a good home, and a good life. But Captain Alexander Johnston, ever since he was a young man, spent his entire life at sea; it was what he knew, and it was how he provided for his family.

But the captain's calling would keep him away from his wife and children; he would miss the best years of the couple's life, and he would not be there to watch their offspring grow. "The Father managed to spend the better part of a year with Mother, following his retirement, and most of that time she was deathly ill," said Sis. The captain's youngest son Neal, a sickly lad since birth, was in the advanced stages of the TB that would also claim his life at a far too early age.

After Nettie's passing in 1928, Alex, Sis and Neal would do the best they could to get by, but they all knew that things would never be the same. The big brick house now had a hollow sound and feeling, Nettie's violin was stuffed into the closet, silent and gathering dust. Alex cooked dinner when he could, Sis took care of the house work between her studies for school and in a few short years, Neal's TB would choke the air from his lungs and take his life.

Shortly after the stock market crash of 1929, the whole world seemed to be in mourning. There was no work, no jobs, and going back to the boats was not an option for the captain. The economic depression was sinking the thin profits of the few boats that were still working. A growing

number of old wooden steamers were rotting at their once
bustling docks, there was no market for iron ore, there were
no factory orders for coal, and by the end of the 1920's the
region's once famous lumber industry vanished. The aging
fleets of wooden steamers that once dotted the Great Lakes
sat idle in troubled water, waiting for a crew that would never
board her deck. Too many of the huge proud boats of steel
were also swept into submission or semi-retirement by the
wave of despair, as America and the world's economy, sank
deeper and deeper into economic oblivion.

Neal Johnston's coughing and hacking became worse, his
throaty coughing only seemed to resonate the rattling of the
heavy fluid and thick mucus that was filling his lungs, stealing
the air he would need to stay alive. Sadly, but luckily for Neal,
his suffering would not go on forever. Neal Johnston succumbed
to advanced TB and passed away quietly, without pain. Neal
Johnston was just 30 years old at the time of his death.

It was thought the youngest of the captain's sons was blessed,
he was not expected to live half as long as he did. Now, that big
old house, once the proud home of the captain and his pretty
wife, was too much, too big for just Alex and his only daughter.
The place where Sis had been born had become nothing more
than a cold empty shell of a house. There were only two of
them now, they did not need such a cavernous place for
themselves, and with the Depression tightening its choke hold
on American prosperity, and Alex's thinning budget, it was
time to move. It was time to find a smaller, more affordable
home, away from the ghosts of his wife and son, away from the
time-muted laughter of sweet memories that were echoing
through the captain's lonely castle.

Alex and his daughter found an apartment on the next
street a few blocks away from their house. "It was smaller
quarters, but had ample room for the two of us, it was a little
closer to town, and more convenient, at least that's what we
told each other," Sis said. Life would never be the same for
either one of them; the change of address would do them

both some good. It would give them both a fresh start, and they would not be constantly reminded of the emptiness in their hearts, by the emptiness of that old house. Alex would do his best to lend some normalcy to their lives, in their new home.

The captain loved to cook and would often work in the kitchen on the culinary creations he learned as a boy working as a second cook on his first boat. His best dish, and his favorite food, according to his daughter was boiled dinner. In the midst of preparing one of his meals, the captain was persuaded to take a moment from his work to pose for his daughter in the driveway at their new apartment.

Captain Johnston on break from KP duty / photo Sis Johnston

Sis would continue her studies to become a teacher. She would also serve as chauffeur to her father, who did not care to learn how to drive an automobile. It was the helmsman man who steered the ship, by the captain's order. "Alex preferred to travel as a passenger in a motorcar, from there he could give commands, on how I should be driving my car," said his daughter.

A sassy Sis Johnston and her wheels /
photo Captain Alexander Johnston (ret.)

Sis loved her father, but her time to take care of his needs and wants was growing shorter and shorter as she juggled his care and her schoolwork and her personal life. Sis would have many important things to take care of leaving her little time to entertain her father with endless card games. Instead she would take the captain down to the Masonic Lodge, "where he could play cards to his hearts content, with the boys, Sis would say. "The lodge was his second home, at least one of his second homes; the Forrest City House Hotel, was also one of Father's favorite stops."

Masonic Temple / courtesy Leon Katzinger collection

The captain achieved his 32nd Degree as a member of the Masons. The captain wears his Masonic charm proudly on his watch chain in a photograph taken to commemorate his accomplishment. (see chapter two for photo)

The Masonic Temple in Bay City, Michigan, stands today as it did during the days Captain Johnston made his daily appearance, as does the Scottish Rite Cathedral.

During the course of researching this project, a (reported) piece of United States colonial history was found tucked away in a quiet corner of the Scottish Rite Cathedral. Encased in glass, and protected from direct sunlight, is an old chair, believed by some, to have played a role in the American Revolution.

It is said to be the chair, once used by the highest-ranking English officer to step foot in the colonies. General Lord Cornwallis, is said to have used the chair, as he signed the instrument of surrender, in the presence General George

Washington, bringing the American Revolutionary War to its conclusion.

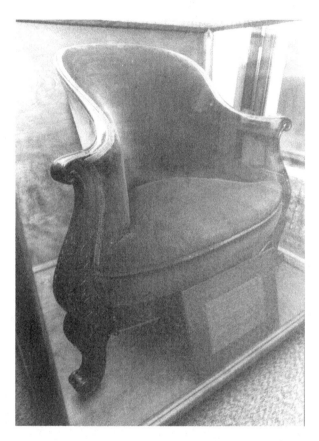

Cornwallis Chair / photo Richard VanNostrand

The red stone of Bay City's Masonic Temple remains as it was, when erected in the late 1800's. A fire in 1903 destroyed the roofline architecture of the building, and its tower. Hundreds of pounds of copper sheeting that once capped the tower and other ornamentation was separated from the debris, re-melted and struck as commemorative coins.

Captain Johnston's original Masonic ring still exists, his name is clearly inscribed with the commemorative date of

January 15, 1894. Johnston would spend his golden years at the Masonic, telling tales of the sea, as he shuffled the deck of cards for yet another game of 500. Alex posed for the last known photograph, of him, taken in mid-December of 1935, at the age of 79 years.

Captain Alexander Johnston (ret.) / photo J. Brandow

The Captain lived out his life under the care of his daughter, Sis. She would make many sacrifices in her life to take care of the father she loved so much, the man who loved her mother with all of his heart. "Father stayed up on current events; he witnessed the beginning of World War II, when Nazi Germany invaded Poland in September of 1939. The captain was fully aware of Japanese aggression and expansionism in China and the South Pacific region. However, Alex would pass away, months before the Japanese attack on Pearl Harbor. But, like all Americans at the time, he knew it wouldn't be long before the war would touch the United States," said his only daughter. Born in neighboring Ontario, Canada. in 1856, the captain had surely seen a lifetime of unending change. The captain's great-great-grandchildren, carry on the Johnston name.

Bay City Times
January 9, 1941

Ex-Great Lakes Captain Passes
Alexander Johnston, 84, lived here 47 Years.

Capt. Alexander Johnston, 84, former Great Lakes skipper, died today at his home, 709 Grant place. He had been ailing for several months and seriously ill for the past two weeks.

Born in Moore, Ontario., Dec. 11, 1856, of Scotch-Irish parents, he came to Bay City about 47-years ago. He was a ship captain for many years until his retirement in 1927.

In 1897, Capt. Johnston was married to the late Nettie Neal, daughter of Capt. William Neal, of West Bay City. Surviving relatives are two sons, Frank B. Johnston of Manhasset, L.I., N.Y., and Charles M. Johnston, of Washington D.C.; a daughter, Alexandra S. Johnston, of Bay City; and three grandchildren, Ruth, Brock, and Heather Johnston, of Washington.

Capt. Johnston was a life member of Joppa Lodge, F. & A.M. Bay City, Chapter, R.A.M., Bay City Commandery, Knight Templar, Bay City, Consistory, A.A.S.R., and Elf Khurafeh Shrine, N.O.M.S. He was also a member of the Ship Masters Association.

Funeral services will be held at 3:30 p.m. Saturday at Hyatt's Chapel. The Reverand Harold McCausland will officiate and burial will be in Elm Lawn cemetery under auspices of the Bay City Commandery.

It was time for Sis to move on. "There was no need to worry about Mother, Father, or brother Neal, their pain and suffering was finally over; they were now with their creator and Savior. Their mortal remains were buried near the chapel at Elm Lawn Cemetery," Sis said. "I know they are in heaven, with God."

The dream of following in her mother's footsteps had come true; for Sis. She had earned her degree, allowing her to become a schoolteacher. Sis's former students say that she was a tough teacher, who gave her students very little rope for horseplay. Some of her students thought Miss Johnston was too strict and at times assigned too much homework. Sis would tell you, if her students had ever met the captain, and had to play by his rules, they would consider her somewhat more lenient.

Sis would say she "didn't care about her popularity, she cared enough about her students, to get them to expect more from themselves. After all, once those students were out in the real world they would learn the true meaning of tough. And, if they could get through her class, there was a good chance, they would make something out of themselves, after graduation.

Alexandra Johnston, taught history, government and physical education. Miss Johnston also spent a great deal of her time coaching women's athletics. Alexandra Johnston, retired from teaching career in 1972, but her quest for knowledge never faded. She remained an avid reader, a lover of documentaries, and kept up on the news events of her community and the world. Sis would also insist, the writer include her status, as one of the biggest fans of the University of Michigan football team, "win or tie." In 1983, Sis Johnston was credited with crossing the Arctic Circle, a promise she made to herself as a child.

Chapter Five

"Uncle Walter" Neal / Iron Man of the Great Lakes!

"Uncle Walter, loved the title, Iron Man," said Sis, "and it was well deserved, at least according to some, but not to his rival, my father. Both Captain Walter Neal and brother-in-law Captain Alexander Johnston, had many things in common, and so many more things, which were not."

"Walter Neal was a rule bender, a risk taker, cussed better than any sailor, on any boat on the Great Lakes, and would push the envelope wherever and whenever possible. Walter also had a most redeeming side to him as well, being somewhat of a kid himself. Walter loved children, and would pay special attention to children everywhere, especially his own and his niece and nephews. Those who have seen the good captain in action say, Captain Neal could tell a salty sailor's joke so well, that everyone who would listen and laugh, would later tell Walter's jokes as if they were their own. The children who were within ear-shot of Uncle Walter thought they were getting away with something; they loved his wild stories, because Uncle Walter's stories were true! Father on the other hand was a rule maker and a rule follower, and certainly a man of Victorian tradition." Alex Johnston's wife, Nettie, and her brother Walter, were modern thinkers, which sometimes put Alex at uncomfortable odds with both his wife and brother-in-law. Alexander, while a strong believer in tradition, may perhaps have been born 200 years too late, give or take a few years either way," recalls Johnston's daughter.

The earliest known photograph of Nettie Neal and Brother Walter was taken in Canada in the late 1860's. "Nettie and Walter had a special bond as sister and brother; some times they had only to look at each other to know what the other was thinking. They were also friends who stayed in touch with each other their entire lives. It was that friendship that also kindled a special relationship between me and Uncle Walter," Sis recalled with a proud smile.

The son of a sailor, Walter Neal would carry on a family tradition that began in England hundreds of years ago. His father, Captain William Neal, first set sail as a boy; he was born in Great Britain and immigrated with his parents to Ontario, Canada, near the Detroit River. "There was plenty of work for sailors, captain or crew," Captain Neal had told his niece.

Historically, the Detroit River has served as a watery battleground for hundreds of years. The native Americans, along with ships of sail, under the flags of France, England, and the blossoming United States, at one time, or another, have battled for control of the valuable seaport and gateway to the Great Lakes. He, who controlled Fort Detroit, controlled the three largest of the Great Lakes.

While only a boy, but within arm's reach of becoming a young man, Walter Roger Neal would carry on the family's legacy. Young Walter would begin his life's work as a merchant sailor, aboard *The David Mills*. She was a relatively new steamer, made of white oak, and ready for her work to begin. Young Walter, was also ready to begin, starting where all young sailors do, in the galley. "You have to crawl before you can walk, he would say," said Sis. "You'll have to peel tons of potatoes, and dump many, many pails of kitchen slop over the rail, before you can even begin to think about working your way, up top, as an able bodied seaman," Sis recalls Uncle Walter saying.

The David Mills / photo courtesy Ralph Roberts

Not only would Walter stay busy as the *Mill*'s official spud peeler, clean up man, and cabin boy, he would also keep his ears clear and wide open so he could learn how to become a sailor. Young Walter would listen to the men as they sat for dinner, or took a sandwich on break. "Walter said he enjoyed hearing the tall tales spun by his older shipmates on the adventures that awaited him, once he filled out," as Sis recalls the story. "Uncle Walter said, the guys on the steamer were very good to him, when he was young, and taught him many more things, than just being a sailor, he didn't go into detail on that one," said his niece.

A seaman's work was tough, and a seaman's job was never done, but there was a benefit no land job ever had. A sailor could eat whatever, and whenever he wanted to, while serving as a crew member. A remaining benefit, that is still in practice, on the sea today. Young Walter's ship was his home, he had a place to work, sleep, eat and to learn, if he wished to one day skipper his own boat.

Father said, "The best new sailors learn from a wellseasoned crew, who learned their skills as boys, who learned from those who came before them."

Walter served aboard the *Mill's* for ten years and obviously had paid attention and listened very closely to his elder boat mates, working his way up to the rank of 1st mate. Walter was now ready to walk in the same steps of his father. Receiving his command as "Captain Walter Neal", his first charge was that of the *George King*.

The George King / photo courtesy Ralph Roberts

Captain Neal, during his career would skipper a string of vessels, which included the *Arizona*. Neal and the crew of the *Arizona* would become a footnote in Maritime history. She was one of the few vessels, to not only survive the Great

Storm of 1905 on Lake Superior, but make it to the port of Duluth, Minnesota, without so much as a scratch. The *Arizona's* crew would bear witness to sunken and damaged vessels on her port and starboard rails, as she steamed her way into port, just a short time after the skies had cleared of churning storm clouds.

The Arizona / photo courtesy Ralph Roberts

Captain Neal's maritime resume is one, that the era's newspapermen, considered impressive. Among his commissions, the *Curry* and the *Sacramento*, built at the Davidson yard on the Saginaw River. The huge oak and iron rudder from the *Sacramento* has been salvaged and is on permanent display on the former site of the Davidson Shipyard, now known as Veterans Memorial Park. The rudder has been placed on the west bank of the Saginaw River, near a small fleet of rotting wooden hulls, that once belonged to the proud and profitable Davidson Shipyard and Fleet.

The Curry / photo courtesy Ralph Roberts

It would be Captain Walter Neal's command of the steamship *Myron*, that would catapult his name into newspaper headlines, across the country, and Canada. That was of course, after being blessed by God, with a miracle. What happened under his command, of the *Myron*, and under his feet, would create his legend.

The Myron / photo courtesy Ralph Roberts

She worked out of Bay City, Michigan, and was owned by Captain O.W. Blodgett. The Myron's old riverside dock area can be found on the east shore of the Saginaw River, just east of where the old Belinda Bridge once stood. Today, the location is a public boat launch owned by the city. The Belinda's, steel beamed, center section would swing open 90-degrees allowing traffic upriver, to load and ship precious cargos of timber and wooden wares to market. Today, rotting posts, in the area of the Myron's former mooring site can be seen near the east span of the Truman Parkway Bridge. The modern four lane draw bridge was built as a replacement for the Belinda Bridge that was too old, too weak, and too dangerously narrow to handle the increasingly beefier post war automobiles coming out of Flint, Lansing, and Detroit, Michigan.

The Myron was a small, but proud wooden steamer, hand made of white oak. The vessel plied commercial waters, from the 1880's, until her deadly and violent demise, in the late fall of 1919. The Myron was like many work boats of her time, and certainly should not be confused with the luxury liners, and cruise ships, that served the privileged rich.

She was built for a heavy load, and she would carry one, all of her days, sometimes towing up to four loaded barges at her stern. Sis said, Uncle Walter would often tell her, a buck was always a buck and for a sailor, at any station, to make any money, they will have to sail rain, shine or other wise. All working sailors knew the seasons last run can be their very best, often with a generous bonus. Captain Walter Neal and his crew, also knew, their last run could be the very worst, there was a reason for such a large bonus, for only a single run.

And sail we shall, with the Myron and the schooner barge Miztec fully loaded, with nearly two million board feet of lumber, the captain and crew departed Munising, Michigan. (on the mid-northern shore of the state's Upper Peninsula)

The *Myron* was headed home, a little over a days sail away, to Bay City, Michigan. Surely the bonus, for bringing home a record load, would pay for the biggest Thanksgiving Dinner a homesick sailor could hope for. For those with wives and children the bonus would take care of a big turkey dinner, with all of the traditional trimmings. There would even be enough money left for the Christmas gifts Saint Nick would be putting under the tree, for the children. At the very least, for those without a home and family, the bonus would pay for a new set of clothes, and a pair of boots. The tighter fisted sailors would have enough money left, to pay rent, on a heated room, and drink a shot of top shelf whiskey, with their beer, for most of the winter."

The Schooner Barge Miztec / photo courtesy Ralph Roberts

"The dark menacing winter skies of the Great North were again threatening the last crews at sea. It was their final run, on the most dangerous of all the lakes, Superior. The biting November wind screamed and howled as it cut deep, past a sailors' bones, and into his soul. Superior's deafening winds warned them, that death was near, it would be a deaf ear all hands would turn. Because the day after tomorrow, they would be home, with a pocket full of money, and time enough to spend it. Unknowingly, back home, at the *Myron's* home port, a sailor's wife's worst nightmare was about to be realized in just a few short hours. The worst, of the worst things that could happen, would unfortunately come true. As the last of his load was put onboard, Captain Walter Neal looked to the cloud filled sky, knowing if he did not cast off, the impending freezing temperatures and the thick ice it delivered, would shackle his boat until the spring, and leave the *Myron* and its crew in irons, far from home. Without a bonus for captain or crew, Uncle Walter told me," said Sis.

The *Myron* had been making her way east, with a pounding North West wind hammering her stern, on the port side. The sea spray coated the chugging wooden steamer with layer after layer of ice. The mounting weight from the thickening ice pushed the wooden workhouse deeper and deeper into the churning waters of Superior. After more than thirty years of faithful service, the *Myron* was about to breathe her last breath. Freighters within sight of the foundering vessel, who were also on their last runs, reported her demise, when the tiny lights of the *Myron* vanished from sight. News of the *Myron* falling victim to the unbridled anger of Superior was telegraphed back to shore. The haunting news spread quickly, and the unfolding tragedy, set newspaper presses across the county and Canada into motion.

Newspaper men had been very busy, for the last year, covering the World War, and the world Flu epidemic that claimed millions of lives, both were finally over. The peace treaty at Versailles, France, had been negotiated at the former home of the Sun King, Louis the XIV. The location served as an exotic backdrop for still and newsreel photographers, capturing on film, the images of the Great War's victors, as they toiled with the language and conditions they were setting for post World War I Europe, and Germany in particular.

Fallout from the Chicago Black Sox baseball scandal was still headline news and fodder for gossip at beer gardens everywhere. No one could figure out why the nation's best ball players would toss away a World Series, along with their careers and their integrity, for a handful of dirty money.

Breaking news, of her sixteen lost crew, and the storm beaten *Myron* would push the headlines of the day, and Chicago's shame aside. Newspapermen jumped on the *Myron's* story. The shipwreck and its sinking had action, high-adventure, mystery, chaos and the struggle of life and death.

Good reporters find great stories, and great stories, sell newspapers. News of the *Myron* and her crews' imminent demise is the stuff, great stories are made of. There was the slimmest of chances, for the *Myron's* crew to survive an icy battle with the Witch of November, but there was hope. The sailors' wives and children, back home, prayed for a miracle. The storm raged on, as rescuers risked their own lives to save the men of the *Myron*. "Newspaper accounts of Uncle Walter and his boat sold like dime novels of the Old West. The sinking of the *Myron* created a maritime hero, and the legend of the 'Iron Man.' It was the perfect story that would sell many newspapers for days to come," the captain's niece recalled.

Only a few hours after her departure from Munising, the *Myron* foundered, off the sailor-respected shore of Michigan's Upper Peninsula. That's when the beginning of the end came for the *Myron* and her crew, near Whitefish Point, in November of 1919. *The Myron*, in just a matter of moments, would come to rest in 60-feet of icy blue water not far from the wreck of the *Edmund Fitzgerald*, lost with all hands, in late November of 1975.

Canadian Singer/Songwriter Gordon Lightfoot immortalized the plight of the *Fitzgerald* and her crew of 29 with his classic ballad more than 56-years after the *Myron's* demise. The lost sailors of both the *Myron* and the *Edmund Fitzgerald*, are brothers of the sea, their voices, along with thousands of others, were forever silenced by the roar of Superior.

It was Captain Walter Neal, from the rowdy lumber town of Bay City, Michigan, whose life, both fate, and the Witch, would choose to spare, becoming the *Myron's* sole survivor. Captain Neal ordered his crew of sixteen into the lifeboats when it became painfully clear they were going down. "I knew we were in trouble, I got the crew into the yawls, and I was ready, come what may!" Sis said, Uncle Walter told her that story for years."

Running rough, in grinding seas, the struggling *Myron* was weighted down by her bonus payload. The *Myron* became the victim of the heavy sea spray that instantly turned a pounding sea into thick layers of deadly ice. The ice grew too heavy to chop loose, it covered her hull, her cabins, and the towering stack of her exposed cargo. Her worn, hand built carcass of oak timbers and planking grew weaker as the storm gathered its might. The *Myron*, began to twist out of shape, as the sea grew taller. The vessel's, creaking and cracking, became louder, and louder, as she rode deeper and deeper into the churning waters of Superior. The weather-worn workhorse moaned as she struggled to inch her way home. The worried coal passers fed the *Myron's* boiler like there was no tomorrow, and there wouldn't be if they stopped shoveling now. Her furnace cooked hotter and hotter with each shovel of coal, that found her chute. The men were keeping up, in the worst of conditions; her pumps however, were not. There was too much water rushing in, and the pumps could not push enough water back out. The *Myron's* coal passers kept shoveling, burning her 700horsepower steam engine to its maximum. Getting the coal to the boiler was in itself a battle, trying to shovel coal into the mouth of the boiler, while being violently tossed about, during one of the worst storms anyone on board could remember. The men prayed for a miracle and hoped this would be the last obstacle of the run, before making it back home to their families. They would certainly have a story to tell.

The crew would be half-right; this would be the very last obstacle of the run, it would be their last run, ever. The rising water was ice cold; it hissed and sizzled as it began hitting the boiler, from below. The men of the *Myron* knew at that instant, what that terrible sound meant, get out and get out now. Captain Neal ordered his crew to the lifeboats. He was going to stay behind, "Come what may." The boiler could blow if the icy waters met the steam chamber, turning the boiler into a bomb. "Putting your ass in freezing water is a hell of a lot better than blowing up," the crusty captain told his niece.

Wrecked on her last run of the season, the *Myron's* crew of sixteen, would eventually freeze to death in their attempt to escape destiny, on that Saturday afternoon, November 22nd, 1919. It would not be until the next week, and then again the following spring, when members of her crew would finally be found. Some of the contorted corpses were encased in the thick shoreline ice of Michigan's Upper Peninsula. The ill-fated crew, their faces, and twisted mustaches, were clearly visible, as their frozen bodies lie suspended in their icy coffins. Then began the grizzly task of thawing the blocks of ice, to free the men, so they may be placed, in a mass grave, at a nearby Indian Cemetery on Mission Hill.

Mass grave of the Myron's unidentified crewmen /
photo Marc Scott

It is a steep ride to visit the crew's mass grave; the narrow single lane road climbs the side of Mission Hill. To some, the ride takes an uncomfortable plane upwards; to others, it is the absence of a guard rail on the climbing hillside road, that may prompt a person to take a tighter grip of the wheel, as they take the white knuckle ride on their skyward pilgrimage. One witness, who saw the men's frozen corpses as they thawed, recalls her childhood memories, of the forbidden sight. "The bodies and mustaches of the men were twisted and contorted as they lie frozen, encased in thick clear ice." The eyewitnesses video taped interview is part of the collection, of the Great Lakes Shipwreck Historical Society, at Sault Ste. Marie, Michigan.

It was Captain Walter Neal who refused to board the yawls, choosing instead to stay with his appointed commission. The captain believed he could steer the foundering *Myron* towards the shore, to ground her, and at least, save her from being abandoned.

As the sinking *Myron* drew deeper into the raging waters of Superior, battered by wind and waves, the once proud vessel, began her death dance, as she twisted and shuttered out of shape. With his crew thought to be out of harm's way, Captain Neal, remained at the helm. As the *Myron's* merciless beating continued, Captain Neal attempted to maneuver his sinking boat's rudder. With the rising water in the wheelhouse already to his knees, Superior raised her icy hand and with a wave of her might, tore the pilothouse from her deck, with the good captain inside.

Captain Neal managed to scramble atop of the wreckage, there he would wait for his maker. It would be up to God, if he were chosen to join his crew and the countless other souls who have been swallowed by Superior over the centuries. But, there was the chance for a miracle, could the captain do the impossible and beat the odds that were so heavily stacked against him? Walter Neal knew it would take nothing short of a miracle, and a little luck, to save his soaked and shivering hide. It was at this point, Sis said, "that Uncle Walter, ascertained for himself, when the last time was, he offered a prayer to the good Lord. But, it wouldn't matter, he would certainly say enough prayers now to catch up for the time he's lost, and perhaps with a few extra prayers, just for good measure," Sis chuckled.

Soaked, battered and frostbitten, Captain Neal managed to hold on to the tiny piece of wreckage for 20 hours in storm churned Superior, until his miraculous rescue by a passing Canadian freighter. The rescue of Captain Walter Neal was the biggest news to hit the papers since the war ended. Canadian and American readers alike love a good story, and they had found one in the rescue of the Iron Man. It was the newspaper reporters of the day, who dubbed the captain the 'Iron Man,' for surviving such an ordeal, which usually, for a mere mortal, resulted in death. Enterprising news writers would work every angle of Captain Neal's story, get it to press and the waiting readers.

Captain Walter Neal's newfound fame and notoriety would also grant him special favor and status other sea captains of the day were sometimes not afforded. While his brother-in-law, Captain Alex Johnston, might describe Captain Neal as a reckless rule-bender, with a penchant for trouble, others would simply call him lucky. "Yes, Uncle Walter was a very lucky man, making it out of more scrapes with destiny and the law, then most," according to his loving niece Sis Johnston.

While many accomplished maritime writers have shared stories of the *Myron's* sinking and Captain Neal's survival, it wasn't until this project, that the personal story of Captain Walter Neal could be told, in its entirety. It was Alexandra Sis Johnston, Captain Neal's niece, who remembers looking Uncle Walter in the eye, as she heard his story, about the *Myron* and her crew, word for word, for more than 30 years, until the time of his death.

The intent, of this project, is to introduce Captain Walter Neal, the man, through the spoken word of the last living person, who heard his voice, and was held in his lap, as he recalled his stories, as if they were told only yesterday.

"Uncle Walter, was big and strong, and would laugh at my jokes, and I would laugh at his," said Sis. As the aging niece recalled Uncle Walter's many stories, the eyes of a child would begin to twinkle, as Sis began to speak. His kindness and his humor lived in the heart of Sis Johnston, until her passing.

Sis never had to worry about Uncle Walter getting out the switch, it was highly likely that it was Uncle Walter who was stirring up all the trouble. "Often times, it was Uncle Walter, giving the children ideas of how to have a little fun, beyond the watchful eye of Mother and Father. "He was a big kid, and I liked that. It was Uncle Walter who always had a piece of candy and a big hug for me. It was also Uncle Walter, although forbidden by his sister, who would slip his special niece a penny or two to spend at the candy store," Sis said.

It is through the eyes, the ears and the heart, of Sis Johnston's memories of Uncle Walter, that we get to know the Iron Man. The man who was her father's not so silent rival, her mother's only brother, and the man who always had a smile, a hug and a few shiny pennies for his "Little Sis."

Chapter Six

"Uncle Walter's Boat has gone down!"

The cold afternoon air, of a late midwest fall, brings with it, a stiff slapping wind and a hard stinging bite. The burnt amber leaves that remain aloft on the huge golden oaks, will stay tight to their limbs for the winter.

In the North, daylight is at a premium; the year-end sun falls quickly, giving way to the cold black night sky. But this is a happy time; the holidays are right around the corner. Halloween has come and gone, with its ghouls and goblins leaving behind waxed windows and smashed pumpkins on the doorstep.

Sis Johnston recalls, "back then, a child would dream up a costume, make it out of rags and other things found in the attic, and go out that night trick or treating to collect a few jawbreakers, licorice and maybe some chewing gum. If we were real lucky we would get an apple or maybe an orange. The rest of the season's holidays, of the Victorian age, were celebrated much the same way, we do today, with a few minor changes. We now have many hours of nationally televised football games to watch, before, during, and after our Thanksgiving dinner. No longer do people spend long hours with their extended families, playing checkers, or cards, or sitting around the parlor to look at pictures from the family photo album. We've replaced the burning candles on the Christmas tree with safer twinkling strings of electric lights. The old timers will tell you that a traditional Thanksgiving would begin with a salute to the pilgrims of Plymouth Rock fame and a prayer. Then, the day was crowned with the traditional turkey dinner, and the special times afterwards where family would spend many hours enjoying each other. Those were very good times when we could all be together," says Sis Johnston. "We played cards and board games, and we would leaf through scrap books containing family treasures like photographs and newspaper clippings and locks of hair."

"In our modern times, it would be difficult to imagine, Thanksgiving without parades and endless football games on a big screen television set. Thanksgiving was a huge event; it was planned weeks ahead of time, and cooked at home. Everything from the pumpkin pie down to all of the fixings; they were made by hand, using family recipes handed down from mother's mother. Even days before the big feast, there was much work to tend to, before anyone would be sitting down, to celebrate the carving of the turkey," Sis said.

Holiday preparations were underway at Captain Alex Johnston's home. The skipper's sturdy two story brick home offered his family shelter from the cold winds that began to slam against his castle. The wind was blowing hard, in sustained hammering gusts. The two oldest Johnston boys, Frank and Charles, ran to the market for a few last minute items that Mother needed for her dinner that evening. "Mother said I would have to stay with her, and work in the kitchen. On Mother's list was a pound of butter, a can of milk and an acorn squash. Mother was a modern thinker but still clung to so many of those things we call old-fashioned today. Mother always used china, when she served her dinners. Mother had a special set of china for our lunches and a set we used for our supper. Mother also had a spare set tucked away, just in case. And thank goodness, just in case, meant we would use the special china for our

Thanksgiving Dinner. But no matter which china Mother had on the table we always used the sugar bowl that she received as a wedding gift when she married Father. A simple but elegant bowl trimmed with gold ivy edges. The delicate handles on either side were trimmed with a touch of blue; Mother said it was the color of Father's eyes, remembers Nettie's daughter.

"On their walk home from the market the boys were already hatching a plan to make sure they would be the lucky ones, who would get the turkey drumsticks, this Thanksgiving. From the wet, and shiny, brick lined street, the boys stopped in their tracks, to admire the first real sign of the holiday, it was their home. A soft, warm, yellow light spilled out onto the ground from every window on the first floor. Sis remembers the boys telling her later, the curtains were open, and inside you could see Mother and I moving about; we were both wearing aprons. The boys looked at each other, and then back at the house, knowing with a few more steps behind them, they would be sinking their teeth into a thick slice of Mother's fresh baked bread. She had pulled it from the oven to cool before the boys left for the market, a few blocks away; surely it was cool enough to slice by now. Now, they had plenty of butter for the warm bread. Maybe later, if Mother were in the mood for a treat, and our chores were done, she would make a batch of popcorn, with hot melted butter and salt. It was the best time of the week for children, it was Saturday night, the chores were done and we had plenty of time to stick our noses into a favorite adventure book, or play a game." They also had an extra bonus coming; it would be a short school week, because next Thursday was Thanksgiving and there was no school on Friday, giving them a rare four-day weekend.

"Mother had just fed a few small pieces of wood into the stove, there was still much baking and cooking to do before dinner would be ready, and the holiday was coming fast. She wiped her hands and picked up the pencil from her recipe desk and marked off another day on the big kitchen calendar. It was November the 22nd, 1919, and according to mother's calendar Thanksgiving was just a couple of X-marks away, Sis said. Mrs. Johnston would often hum and sing songs while she was working in the kitchen. On Thanksgiving the house would be filled with a most wonderful aroma of oranges, pumpkin pie and fresh baked bread. No matter where you were in the captain's huge house, your nose gave you a tantalizing sneak preview of what the taste buds would soon enjoy. We were told to stay out of Mother's way while she worked; young faces would just happen by the kitchen door just in case a bowl or a mixing spoon was to be licked, or a small sample was to be tested, before dinner." Maybe this year Frank, her oldest son, now in his mid-teens, would be carving his first turkey, if his father didn't make it home.

Captain Alexander Johnston didn't know if he would be able to get home; there were still ships to inspect, and logs to audit, on the season's cargo and a total equipment inventory of the fleet before his work was done for the year. "There were a few years when there was a labor strike and Father would have to spend the entire winter season guarding the ships at the dock. Sometimes he wouldn't make it home that year," his daughter recalls. But this year, the captain would try to make it home, his work would take several more weeks, and once his task was finished he would board the train for home. Father said, "he would call and let us know, one way or the other, if he could get to a phone." Not one of the three Johnston boys would have a chance to win the race with their younger sister to answer the telephone. Sis seemed to have a knack for knowing when it would ring; she was always the first one there, to answer it. Sis seemed to enjoy that certain pleasure, beating her big brothers to the call. Sis was hoping to be the one to take the call from Father, saying he's coming home, because she wanted to be the one, to tell Mother the news.

Unfortunately it would be Sis, to take the phone call, that would rip their hearts out, and turn the Johnston/Neal family upside down with shock and disbelief. While waiting for a call from Father, the phone rang, and again it was Sis who got to the telephone first. "Mother, it's Aunt Maggie!" In the time it took Nettie to walk the few steps from her busy kitchen, to the library, a silent alarm began to pound in the hearts of every one in the house. Even the boys upstairs, felt they needed to be downstairs, at that moment, for some strange reason. A Victorian Thanksgiving, with all of the trimmings, was a highly celebrated occasion in the Johnston house. "Before Mother put the phone to her ear, we all knew something was terribly wrong. Everyone seemed to already know that one of our most anticipated family holidays would soon turn into the worst the family has ever known."

Sis looked me straight in the eye and fixed her gaze; it was like she was looking right through me. As I met her gaze, a strange, but comfortable, far away look, overcame her, and she began to speak. With each word she uttered, we took giant steps, back in time, to Saturday, November the 22nd, 1919. While this may sound somewhat odd, Sis seemed to enter a trance-like state, as she recalled that day, and its events.

"The telephone rang, on the other end was Aunt Maggie, and she did not sound like herself. Aunt Maggie had only to speak a few words and I knew something was wrong, terribly wrong. 'Put your mother on the phone!' This was not like Aunt Maggie, there was no hello, no how are you, no chit chatting or small talk. Aunt Maggie was a funny person, with a great sense of humor, but not on this call. Put your mother on the line, please! I called Mother into the library from the kitchen."

"Mother put the phone to her ear, and said only one word to Maggie, Yes?. Mother stood rigid and erect as she took Aunt Maggie's call on our silver candlestick phone. The conversation lasted only a few moments, but it seemed to last an eternity. Mother remained calm and collected as she placed the earpiece back on the hook and stood completely still for a moment, not wanting to believe what she had just been told. Mother then put the phone back on the top of Father's big roll top desk."

She turned, to find all four of her children, waiting very quietly, for her to share the news, from Aunt Maggie. "Mother, what's wrong, what has happened? "Nettie solemnly summoned her children closer, to sit with each other on the couch in the library." She looked us in the eye and said, 'Uncle Walter's boat has gone down and that's all we know right now. Children, we all know this is something that could happen, at any time, to any sailor, whether it be Father, or Uncle Walter.'"

"We don't know what has happened to Uncle Walter, or his crew, but Aunt Maggie says there are men out searching for his crew and his boat, right now. We may not hear another word of it, until well after daybreak tomorrow, or longer; all we can do now for Uncle Walter, his men, and those people out there trying to find them, is pray."

"Uncle Walter" / courtesy Great Lakes Lore Museum

How right Nettie was, there wasn't anything, anyone could do; not even those men, who witnessed the heartbreaking tragedy, of the doomed vessel, and her dying men. They could only watch, from the rail, as the men screamed helplessly, through the roar of the sea. The doomed men were just out of reach, but close enough, for their disbelieving eyes to witness.

Sis recalls Mother being especially kind with her words, as she broke the news of the *Myron's* tragedy. *The Myron's* captain was her only brother, to her children, he was their favorite uncle. "Mother told us that was all she knew, and not to bring it up, unless she asked them to. It could only be hoped, that it was the darkness of night, that would keep the rescuers, from spotting Uncle Walter, and the *Myron's* crew. "It was the hope, that daylight would bring the crew's safe return, if they could only last the night and survive the beating. There was hope, perhaps not much of it, but there was, at least for now, the uneasy comfort, of the slightest chance, for Uncle Walter and his crew's survival.

"Mother sat quietly with us for only a moment, she was calm, but we were stopped in our tracks with the news."

"We were all in the library and then went off to our places in the house. The boys, went back up to their rooms, it was too cold to play outside. I went into the kitchen to help Mother, after all, dinner was still cooking and Thanksgiving was only a few days away, and there was plenty of work to do, in that kitchen. From that moment forward, the relationship between my mother and I was different. We weren't just mother and daughter; we were also very dear friends. We both knew, we had to be strong, we knew everything was in God's hands, and we knew we had to carry on."

"Mother loved her only brother very much, and I loved my Uncle Walter, even if Father, had little to say that was good about his brother-in-law. Father thought he was a show off, and a marginal Lakes pilot, at the very least. Father was right about one thing, Uncle Walter, was always in some sort of trouble." For Sis Johnston, Uncle Walter, was one of the brightest spots in her life.

"While Mother and I worked in the kitchen, Mother shared her adult thoughts, in a soft voice, while we worked. Aunt Maggie got a telegram; Aunt Maggie hated the Western Union man, because she knew, it would be, one of those guys, who would deliver bad news. Maggie always seemed to think a hand delivered telegram would be Walter's death notice. Mother was a strong woman, much like that of a soldier or police officer's wife, who knew there could come a time, when the Western Union man could be knocking. We all hoped we would never hear the ringing bell of the telegram man or see his bicycle parked at our front steps."

Across town, on the other side of the steamer jammed Saginaw River, Maggie Neal and her son Bill were in shock; they both sat in silence, at the kitchen table. There was nothing more they could do, there was nowhere they could go, they could do nothing more, than hope for the best, and pray for a miracle. "It didn't seem much like a holiday now, Uncle Walter was in trouble, and all of us feared the very worst. How would we make it through Thanksgiving, what would we have to be thankful for? How could it ever be a Merry Christmas without Uncle Walter? And, what about the men, and their families, who are praying for the same miracle."

The ringing telephone at the Johnston house had everyone jumping out of their skin. "It was Mother who answered the phone this time; I certainly didn't want to hear any more bad news," said Sis. "Mother was only on the line for a moment, it was Father calling to see if we had heard anything new. News of a captain and crew going down flew like the wind, and even Father said he was praying for his brother-in-law's safe return."

"Father said, he wasn't sure if he would be able to catch the train from Cleveland, to make it home for Thanksgiving, but was going to try."

The house lights, burned through the night, at Captain Neal's home, on the city's west side. "Aunt Maggie kept her chin high, during her stoic vigil, waiting for word of her husband's fate. Her only son, Bill, was at her side. Only a year ago, Aunt Maggie and Uncle Walter, prayed for his safe return from World War I. Bill was still recovering, from a painful case of trench foot he had brought home, from the flooded, rat-infested, trenches of the French battlefields. After reading newspaper accounts, of the war, over there, while Bill was in the service, Maggie would keep with an old family tradition, lighting a candle, and placing it, in the front window. It was a flame of hope, that would light the way, for the safe return, of their men at sea."

"Each and every night, after Bill shipped out to serve in France, Maggie would light the way for her son's safe return with a candle in the window. But, tonight it was Bill, who placed a fresh candle in the same window. He lit the flame, with a match, and said a prayer for his father. Bill had been busy working his new job, as a junior manager, at the Chevrolet plant in Bay City. Bill said, one day he would be running the place, and no one doubted his dream. Bill was the first Neal in generations not to become a sailor, and tonight's grief only seemed to underscore his decision. He pressed the burning candle deeper into its simple brass holder, he moved it slightly, from side to side, to make sure it was secure in its standard, standing straight and tall. Although he was not superstitious, and said to be quite the businessman, Bill wanted to do everything he could to ensure his father's safety. Bill was making sure the single flamed beacon would light the path, for his fathers return.

Across town, at the Johnston house, the lights were on, late into the night. Nettie Johnston awaited news, of her only brother's fate. Sis and her family could do little more than pray and keep silent. "Mother had a tremendous burden to carry and she didn't want to be pestered with children's questions about Uncle Walter. Children were to be seen and not heard; it was a rule of the times, and certainly, a rule that was not to be broken, at least not tonight. We all prayed with all of our might, and we were ready, to scamper, when the telephone rang."

"Mother had reminded us, that it was 23 years prior, that our grandfather, Captain William Neal, had died, just after his last run, of the season in 1896, in Ashtabula, Ohio." Now, his only son, Walter, was in trouble, and Mother was left to wonder, if there would even be a body for the family to bury.

While a miracle was hoped and prayed for, little did Nettie, Maggie, or the children know, their prayers were being heard, above the roar, of the raging sea. For during this time, a few hundred miles to the north, in pounding seas, Uncle Walter was wet and freezing cold, but he was still alive, clinging to what was left of the *Myron's* pilothouse.

What does a man, shipwrecked, and alone in rough seas, think about, besides staying alive? "He thinks about dinner; a hot, tasty meal, of his wife's home cooking, dished up from a steaming pot, and served at his own kitchen table. That's what a shipwrecked sailor thinks about," says his niece, Sis.

During the course of our many interviews, over the years, I had approached Sis with what I thought might be some very delicate and personal questions. I, did not know, if I would be invading some her most personal and perhaps most protected thoughts.

Sis, was the last eyewitness, to an era gone by, I had to find out, what she remembered. I asked her, "did Uncle Walter ever talk about the Myron, after his terrible experience?" Sis turned to me, with a grin and a wink, and said, "he never stopped talking about it, never! That's all he talked about was the *Myron*, and the SOB that would not throw him a line. Uncle Walter would repeat the story for anyone, willing to buy him a shot, and sit still long enough to hear him tell the story."

I asked, "how many times did you hear him tell the story?" This time, with a chuckle, a twinkle in her eye, and flashing a big smile, Sis said, "I've heard the story, from Uncle Walter, for more than 30 years. I loved to hear him tell it, and he loved it, when I would ask him."

All the while, Aunt Maggie, Nettie, and the children prayed, time stopped. Sis said, "it would be another whole day, before we heard any sort of news, about Uncle Walter, or his crew, and we were scared."

I asked her, "did he ever tell you what he did for 20 hours to hang on, to not give up hope?" "Well yes, he was mad at the captain of the *McIntosh*, it came so close to him, and the captain would not throw him a line. Uncle Walter, got that captain's license taken away, for it."

"Besides that, was there anything else, that Uncle Walter had talked about, that kept his mind occupied for the 20 hours he was floating at sea?" I inquired. "Uncle Walter is still mad at that captain, even today, if he were alive, he would be cussing out that captain, who didn't stop to help him." Uncle Walter said, he kept looking at his watch, wondering what Maggie was up to back home." It was Saturday, and Saturdays were for shopping and visiting, and Aunt Maggie loved to shop, grocery shopping, shoe shopping, every kind of shopping. Saturday was a treat for shoppers, on occasion, a few of the dry goods stores were open until 9 o'clock at night. With the extra shopping hours, you could make sure you had what you needed for a special Sunday, because nothing was open on Sunday, it was the Sabbath. Uncle Walter knew that his beloved bride Maggie would make the very best of Saturday night shopping, even if she didn't need to buy anything. Shopping was an opportunity, for Aunt Maggie to catch up on the latest gossip. She would bump into a few of her friends and neighbors, who were also out shopping. Aunt Maggie would seldom return home early, if there was a good story to be heard."

Captain Walter Neal went over and over, in his mind, how the storm smashed his boat, and sent it to the bottom of the sea. Captain Neal was still in command of the *Myron's* wheelhouse, but his ship, and her cargo were little more than shattered wreckage, being violently tossed about, in the icy waters of Superior. Numb, from the freezing cold wind and water, Captain Neal, tried to use the fading daylight to peer through the blinding snow squall and search the horizon for the lifeboats he had ordered his men into, before the *Myron* went down.

In an effort to gather his wits, and to get a time fix, for the accident report, he hoped to live long enough to fill out, Captain Neal reached for his pocket watch again. Had it not been for its gold chain he received from his beloved bride, the captain would have lost the timepiece given to him by Aunt Maggie years ago. The pounding waves had knocked the captain's timepiece from his hand, leaving it to dangle for a moment, much like the captain's life, as he clung to what was left of the *Myron*.

This was not the first time an angry sea had threatened to take the life of the aging skipper, but it was certainly the most recent. Shivering and cold, Walter Neal knew, that at least he was still alive, and his life safe, for the time being. Using his pocket watch to connect himself, mentally, to his wife, he took a good long look at the face and hands and began to think about Maggie. Sis said she, "always enjoyed the part of the story, when Uncle Walter was trying to guess what Aunt Maggie was up to back home." "Well, Maggie might be at the market right now getting the turkey and the trimmings for our Thanksgiving Dinner, Uncle Walter would tell me," said Sis. "He would look at his watch a few moments later and wonder if his family back home knew of his predicament. Uncle Walter knew that Aunt Maggie hated telegrams. Should worse come to worse, Maggie was going to get that telegram, on news of the *Myron*, if she hadn't already. Uncle Walter hoped Maggie would not find out about what happened because he didn't want her to worry about him. But on the other hand, he had hoped, that Maggie did know what sort of predicament he was in. She could help him pray for the miracle that he needed, right now! Uncle Walter knew, that men don't usually survive very long in freezing water; he knew he needed a miracle, and he knew he needed all the help he could get, even if he would never admit it to another soul."

"Maggie was a tough old bird, but don't let her hear you say that," said her niece. She had been through many tough times with Uncle Walter. He had been through many storms before, like the big one, on Superior in 1905, but Walter always made it home. And that's when Aunt Maggie would have to go to work on Uncle Walter, when he got home. Maggie loved her husband very much, but it was the love of her life that also caused her many moments of frustration once he walked in the door, for his winter break."

"Maggie, would always say, it took her a few weeks, to break Walter, of one particular bad habit, once he got back home. Maggie wanted Uncle Walter to get in the bathtub at least every other day. Uncle Walter was used to being at sea, where how often one bathed didn't really matter. Or at least, there was not one crew member who would complain out loud about the captain's odoriferous demeanor, since they were all literally in the same boat, and all stinking like dead fish and weeks old body odor. It was Father who would always use the word, crusty, when he would talk about his brother-in-law."

"Uncle Walter had a firm command of his adverbs and adjectives, and was not afraid to use them. Father was also a sailor, and a brilliant man, but did not have the flair for the use of those colorful expressions, the way that Uncle Walter did," Sis said with a muffled giggle. "Still clinging to life and what was left of the *Myron's* pilothouse, Uncle Walter kept looking at his watch, when daylight had faded, he had to put it up close to his face, touching it to the tip of his nose, so he could see the numbers and the hands, it was so dark."

Sis remembers her uncle telling her, "once the waves shortened up it was easier to see, but he was still being tossed around, pretty good. Uncle Walter said, he would look at the time and then close his eyes, and he could smell his wife's kitchen. He dreamed of having a pipe smoke, at the table, as Maggie stirred the broth. before dinner was served. Uncle Walter always thought of himself as a great cook, and he was. His beef stew was the best, of the best, but when he was home, he would defer the task to Maggie. He could sit near the stove, with his pipe, and put his stocking feet on the sleeping dog, under the kitchen table."

"Soaked, shipwrecked, cold, and alone, Walter Neal continued to cling to the bobbing wreckage of what was left of the *Myron*. The shipwrecked captain, laughed out loud, as the minute hand was straight up at twelve. "It's past nine o'clock and my dear Maggie should have her arms full of packages, and be making her way through the front door, after spending all of my money. If my Maggie bought a pair of shoes, like I know she has, she'll have them back on Monday saying she can't wear them because of her feet. Maggie may walk through the house once, or twice, with her new shoes on, but then she will put them back in the box, and leave the lid off. She can walk past them, and curse both the shoes, and her feet, on her way past the dining room table, where she left them. That's just how Aunt Maggie is, Uncle Walter would say with a chuckle."

Captain Neal, wanting to believe his crew of 16 were safely away, in the lifeboats, willed himself to hang on. Not only was his mind on Maggie's home cooking, and how much money, she blew on her shopping trip, he was going to "kick the ass" of the captain of the *McIntosh*, who left him, to die. Uncle Walter said, shortly after the *Myron* went down, the *McIntosh* was close enough to throw a line, but her captain, within shouting distance, said he would send back a tug to get him instead. Uncle Walter said he could read the nameplate of the boat as is went by him, but they would not throw him a line." The northwest winds, continued to hammer on Captain Neal, as he clung to the wreckage; and for a time exhaustion would force his eyes to close. The icy waters of Lake Superior, would force them to open again. Captain Neal had an excellent sense for dead reckoning, he knew he was far from the spot, where the *Myron* went down. If there were crews out looking for him and his men, they were over the horizon, and a long way from his position. Twenty long, flesh biting hours would go by before Captain Neal would finally be rescued from the choppy waters of Lake Superior."

Below is the clip Walter Neal tore from the newspaper in Port Arthur, Ontario, Canada, reporting his rescue. Neal was still aboard the *Franz*, warming up, when he read these words, for the very first time.

Sis said that, "Walter liked to read about himself in the newspapers. But, he would just roll his eyes, and say, they couldn't spell my goddamn name right." The captain had re-read the article recounting his ordeal several more times, before he tore the story out of the "News Chronicle," and stuffed the clipping into his coat pocket. Captain Neal, read the clip, over and over again, on the train home, with Maggie and Bill, who also took turns reading it. It was his niece Sis Johnston, who pulled the clip from Uncle Walter's coat pocket when he got off the train and took it home to paste into her scrapbook.

November 24, 1919
News Chronicle
Port Arthur, Ontario, Canada

EXTRA! Captain of Myron Reaches Pt. Arthur

> *Shipwrecked Sailor, Apparently Only Survivor Among Crew of Eighteen, Is Picked Up By Steamer W.C. Franz Twenty Miles From Where His Boat Went Down After Drifting On Wave Washed Pilot House For Twenty hours.*
>
> *Captain Neal, of the steamer Myron, which foundered off Whitefish point, Saturday, in a gale, as reported in the*

telegraphic dispatches published in this paper, was rescued yesterday morning by a boat crew from the steamer W.C. Franz. Captain W.C. Jordan, which arrived in Port Arthur late this afternoon.

Captain Neal, when observed, was floating on the Myron's pilothouse off Parisian Island, a distance of twenty miles from the scene of the disaster. The wreckage was noticed by Captain Jordan, who turned his glasses on it, but could not at first make out the shipwrecked sailor.

"By The Merest Chance"

"It was by the merest chance that I saw him move his hand," Captain Jordan said to the News-Chronicle, this afternoon. "We then sent out a boat and it was not a difficult matter to get to the wreckage, although when he was reached, Captain Neal was found to be in a very battered condition."

After being taken aboard the Franz the captain was put to bed and given such other attention as could be afforded by the crew of the Franz. It was found that, while suffering severely from bruises, particularly about the legs and ankles, and exposure he had not, apparently, suffered much from frostbites. He was still in bed when the Franz tied up at the Davidson and Smith elevator this afternoon.

Another Boat Refuses Help

Captain Neal, whose home is at Bay City, told Captain Jordan the before his arrival, another steamer passed very close and that he made an appeal to it for help. The reply was, "I'll send a tug after you." Such a tug could not have gotten to the scene inside of ten hours. When the News-Chronicle asked Captain Jordan for the name of the vessel, he declined to give it, although he said it was in the possession of Captain Neal, and intimated that it would, probably be given to the proper officials when the time came to make such a report of the case. Captain Neal says the other steamer could easily have picked him up.

After rescuing Captain Neal the Franz looked about for other wreckage or sailors but could see none and continued her trip to Port Arthur.

EIGHTEEN PERSONS GO DOWN WITH THE SHIP; GRAVEYARD OF LAKES CLAIMS ANOTHER

Men In Life Boats; Seas too heavy to permit rescue

Other Boats See Men Struggling In Water/Bay is strewn with wreckage-Coast Guard Seek Survivors

Soo, Mich. Nov. 24-After a thirty-six hour search, no trace has been found today of any of the crew of the steamer Myron, which went down off Whitefish Point during a storm Saturday.

Hope that at least members of the crew eighteen of the lumber laden steamer Myron, owned by C.W. Blodgett, Bay City which foundered off Whitefish point in a terrible gale which swept Lake Superior Saturday and yesterday, would be rescued, was held out today by marine men here and captains and steamers who passed this port. Boats that ventured down from Whitefish shelter yesterday reported seeing men clinging to her wreckage or lashed to rafts from the Myron, although attempts to rescue the men were futile. It is believe that some of the crew who thus got away from the foundering steamer might drift ashore and be revived by patrol coast guards, despite their long exposure and buffeting by the great waves. Wreckage from the Myron continues this morning to come ashore in great quantities, including parts of the cabins and upper works, while the entire bay is strewn with the lumber cargo of the boat and her consort the Miztec, which was reported towed into shelter by an unidentified steamer.

November 25, 1919
Daily News-Chronicle
Port Arthur, Canada

Steamer McIntosh Sails by Leaving Captain of Another
Boat on Wreckage to Fate

So Relates Master of Ill Fated Myron, Later Picked Up by
Franz and Brought to Port Arthur

"*Some of the McIntosh's crew shouted to me that he*
would send me a tug, but I called back that all I wanted was
a line. I might have been bruised in being pulled aboard but
I would have been rescued. He could have cast me a line
easily enough for the seas were not running high," said
Captain W. R. Neill, of the lost steamer Myron in telling
a News-Chronicle reporter of the approach and passing
of a steamer which sighted him floating in mid-lake on
the pilot house of his ship." The captain of that ship,
the McIntosh, an American boat, would have taken no
risk in rescuing me. To Captain W.C. Jordan of this
ship and his crew, I owe my life. I shall never, during the
rest of my life, forget the many kindness', which he and
every man of his crew, down to the cook, have shown me.
This boat has been my home since I was rescued on Sunday
and will I ever have occasion to think of the streamer W.C.
Franz and her master."

A Regular Old Sea Dog

The News-Chronicle reporter found Captain Neill, in a
bunk in the bow of the ship. He had been in that bunk since
being picked up Sunday noon, 20-miles from where his ship
foundered. A corn cob pipe was in his mouth and on the
table near him was a can of tobacco.

His face was screened by a curtain at the end of the
bunk and in order to get a view of the mariner, the reporter

pulled the curtain aside. Surprise came over the face of the newspaper man on beholding that the captain's hair was gray. "Why captain, you are an elderly man, your hair is gray. You have had a trying experience.

"Yes, I am getting up, I am passed fifty-five years, but I am still alive and with that joy the news of my rescue will be received by my wife at our home in Bay City. You know my wife has always had a dread of telegrams, but she is due to get a good one soon, but it contains good news this time. Captain Jordan sent it away for me as soon as his ship reached here this afternoon.

When I'm home the wife insists that I take a bath at least every other day. I am quite content to take one every week. This is the telegram I sent her today.

"I have been picked up by the steamer W.C. Franz, after twenty hour bath, middle of Superior-am fine."

Tells his story

"Now go ahead Captain tell us the story, in your own way," asked the newspaperman.

Captain Neill took the corn cob pipe out of his mouth rubbed his hand through his hair and thought for a moment. "I don't care to say very much this, the only thing I can think of now is the kindness of the Jordan's crew. I could say a whole lot about that, but you can do it for me.

"We left Munising, Michigan, at 4 o'clock on Saturday morning for Bay City, Michigan. We were loaded with lumber and my ship had in tow the barge Miztec, also carrying lumber. On Saturday afternoon about 2:20 o'clock we lost our tow, the barge having snapped her cables in the storm, a nor'wester which was raging at the time. At 4 o'clock the Myron went from under us and foundered. We kept her going until the water reached the fire in the furnaces, when they went and we had to give up. The crew made for the yawl boats while I remained in the pilot house. I was determined to stay by the ship, come what may.

Climbs on Pilot House

"When the vessel had settled to such and extent that I was standing in some three feet of water, I climbed upon the top of the pilot house and remained until it was knocked off and I went with it. It was growing dusk then. All night long I remained on the pilot house tossed about in the sea, and through it all, I never lost consciousness. I never saw any member of my crew after they took to the boats and whether they got away from the craft or not I cannot say and I may be the only survivor of the disaster.

There were seventeen of us in the crew, including myself. I hung onto the pilot house for dear life. I had been holding onto the fog bell, but when that went over I had to grab the coping. Naturally I had no food with me and I slacked my thirst by dipping up handsful of water time after time. As the night wore on and the sea calmed down, I believe I went to sleep. I suffered no great exposure, though the water was dreadfully cold, my hands and feet were slightly frostbitten and my legs are badly bruised. When morning came I began to scan the waste of waters for a ship that might come to my rescue. I was quite cold by this time and I thought that I'd sooner be in a warm cabin than out in the middle of Superior, so I began to wave my cap frantically.

McIntosh Comes and Passes

A steamer by the name of McIntosh, an American, I believe, passed me, not very far away. I appealed to her for help, and the reply came back, "I'll send a tug for you". They could have easily thrown a line as I wanted only that.

Well on in the forenoon another vessel passed me, but she was some distance away. I kept waving my cap and I thought I had heard her whistle in answer to my signal, but she never came close enough to investigate.

Then about noon on Sunday, I sighted another ship. This time, I stood up to wave my cap. They evidently saw my signal for a few minutes later a boat came towards me and in no time I was aboard this ship which I have come to learn is the W. C. Franz. I was bustled into bed where you see me and everything possible has been done for me since my rescue.

I have already wired my wife that I am safe, wired the owner of the Myron, that she is lost and wired my son, who is the assistant manager of the Chevrolet Motor Works in Bay City to meet me at the Soo.

I only hope that some poor fellows of my crew were as lucky as I have been. I stayed at my post, if not in it on top of it, with a laugh. I will return with the Franz to the Soo and will then go to Bay City, where I live, when I am home.

THROW OUT A LIFE LINE

It seem almost incredible that ships would pass a shipwrecked sailor clinging to wreckage far from shore without at least trying to secure him. That is what two ships did to the captain of the foundering vessel, Myron.

Clinging to the pilot house of his lost ship Captain Neill sighted two ships, one close enough to hail him and the other at a distance and had a sickening experience of knowing that he had been seen and deliberately abandoned to his fate.

This is not the first time that such a thing has happened. Possibly it may not be the last time, but it should be the very last time that any lake captain would do such a thing and still remain in charge of a ship.

What explanation can be given for such callous disregard of human life? The authorities should call the captains of the two ships and demand an explanation.

It is very true that the captain of a ship has his duty to perform to his owners, but should it ever happen that an owner censored or punished his captain for going to the rescue of a life in danger, public opinion would drive that owner off the lakes and cause his ship to rot at its dock for want of business.

There is such a thing as having to abandon men in peril on the high seas because of imminent danger of losing other lives in attempting rescue, but that risk was not present in the case of Captain Neill. Truly he said to a reporter" "We used to have wooden ships and iron men, but now we have iron ships and wooden men. "

November 28, 1919
Daily News-Chronicle
Port Arthur, Ontario, Canada

CAPT. W. R. NEAL, AT SOO ENROUTE HOME, TELLS OF ALLEGED CRIME OF LAKES

Charging that the captain of the steamer McIntosh deliberately and without any need, left him and the remainder of the crew of the sunken steamer Myron on to a fate which meant death in all cases but his own and thereby committed a crime than there has been no greater in the code of the seas, Capt. W. R. Neal told a story of horror which in not equaled in the annals of winter shipwrecks on Lake Superior. I was clinging t the roof of the pilot house when the McIntosh sailed by shortly after the Myron went down from under me, " said Capt. Neal. "The McIntosh drew along side me, not more than sixteen feet away. Although it was dusk their ship was so close I had no difficulty in making out her name. I talked to the captain and expected he would put out a yawl and pick me up but he did not do so, not attempt in any way to help me. "I will have a boat sent for you," the captain of the McIntosh called, and then drew away. I have never seen him since not do I ever want to see him again. "

Sis and her brothers would read and re-read the Canadian newspaper clips of Uncle Walter, and those that followed, many, many times out loud. "Last week, our hearts were torn out of our chests, with word of Uncle Walter's ship being lost. Today, we can't wipe the smiles from our faces." The children were told, to find something to do, upstairs, as Father and Uncle Walter talked about the ordeal Sis said, "upstairs, Frank passed Uncle Walter's newspaper clip to me, for my turn to read again." The children, kept smiling at each other, as they heard Uncle Walter's tired, ornery, but comforting voice, rumbling from downstairs.

"We could hear Uncle Walter swearing louder and louder as he shared his shipwreck story with Father, 'that Son of a Bitch from the *McIntosh* left me out there to die!' Goddamn it, I asked for a line, throw me a damn line! I could see the SOB's face, I could read the name of the ship, it wasn't anymore than 15 or 16 feet away from me. Uncle Walter would say, he said he would send a tug for me, the son of a bitch. Uncle Walter, stayed at the house for a little while, as he and Father shared a few glasses of Scotch. Uncle Walter, only pulled his smoking pipe, from his teeth, long enough to take another sip, from his short, heavy, glass. We all knew, what the men were up to, because we could hear the bottle, clinking on the glasses. Mother said, that we weren't to talk about any drinking in the house, because it wasn't anyone's business, what was going on in this house. Mother knew, we didn't have to be told, but Father, wanted to make sure, the children knew the rules, and we did!"

"My brothers and I all got to say goodbye to Aunt Maggie, and Uncle Walter, before cousin Bill drove them back home. Earlier," Sis said, "I heard Aunt Maggie tell Mother about the big cuts and bruises that Uncle Walter had all over his legs. Aunt Maggie said that there was almost nothing left, of Uncle Walter's pants, when he was rescued. And she said, that he was beat up pretty good out there. Uncle Walter is my hero, and I just couldn't imagine my hero getting hurt, he pulled up his pant leg to show Mother his cuts, and bruises, but I wouldn't look."

"It certainly was a rare sight, I saw Father and Uncle Walter shaking hands and wishing each other a Happy Thanksgiving. That was the first time I ever saw the two of them shake hands, let alone be in the same room together without disagreeing about something. Uncle Walter shook hands with my brothers too, and then he smiled and walked over to me," Sis recalls.

"Uncle Walter groaned a little when he reached down to pick me up. Mother and Aunt Maggie told him to put me down but Uncle Walter just squeezed me harder. He buried his big cold nose on the side of my neck, and then began to whisper in my ear. He thanked me, for the prayers I said for him, when he was out there. And he said, that he thought about my big smile and my hugs, and that helped to save him. I know, my hero didn't want me to see it, but I saw Uncle Walter wipe a tear from his eye, before he put me down and walked slowly out the front door. Mother had already walked Aunt Maggie out to the waiting car; Bill was holding Uncle Walter's arm on the way down the front steps."

"Father, turned to me and asked, if I would call him a cab. I knew, he was going to play cards, at the Masonic lodge. I knew that Father couldn't let a day pass by without playing cards. But, I also knew he really wanted to share Uncle Walter's story with his lodge members, and the other captains who were home for the winter. After all, it is the holiday season, and this fresh news, may prompt a short break in the gentlemen's card game for a stroll down to the Forrest Hotel for a toast to Captain Walter Neal."

"Although it did seem odd, but not unusual, Uncle Walter wasn't invited to the meeting," Sis said, with a snicker. "It was one thing to share a drink, and celebrate his brother-in-law's safe return, in the privacy of his own home, it was quite another to been seen in public with the man," said Father. "It didn't matter to me; my Uncle Walter was bruised, and battered, but he was home safe," said his niece.

Chapter Seven

Back Home

Captain Walter Neal smiled slowly and then chuckled, as he slapped his knee and looked over to his son Bill, behind the wheel of one of the finest 1919 model Chevrolets on the road. Walter Neal didn't say a word, but fixed his gaze on his son's profile. Bill was heading west on potholed and bumpy Third Street, steering clear of the trolley tracks that ran down the middle of the road, thankful for the miracle of his father's survival. Bill was also thankful, as were his other passengers, that he could see clearly enough to commandeer his company car home without crumpling the fender. Bill Neal, assistant manager of the General Motors Chevrolet plant in Bay City, Michigan, was feeling his oats after a few sips at his Uncle Alex's house. It's not often that a person can enjoy the celebration of another person's homecoming, especially after they were thought to be dead! It was one of Captain Johnston's lodge brothers, who also happened to be a police officer that brought over one of the finest bottles of "nerve tonic" a fellow could find, to help celebrate Uncle Walter's homecoming.

Bill drove with both hands on the wheel and his stare, straight ahead as he wore one of his biggest smiles on his face since he came home from the war in France, a year ago.

William James Neal, World War One, France,
1917-18. / photo courtesy Anne Sullivan

Bill drove their Chevy across the ironed-beamed Third
Street bridge, over the quiet Saginaw River, to the west side
of town. Captain Neal, still staring at the silhouette of his
son's profile in the afternoon sun, began to laugh out loud;

Bill not taking his hands from the wheel or his sight from the road also began a hardy laugh and Maggie, who was in the back seat was laughing louder and harder than the two men in the front seat.

But as quickly as the boisterous roar began, it ended. There was a solemn silence that over took the motor car. Maggie, Bill and Walter had all come to realize at the very same moment, that their prayers had been answered, and here was Walter, breathing, alive, and safe in the car heading for home. They all looked at each other, and with a successive nod, acknowledged the happy tears that streamed down their faces. Luckily Aunt Maggie was prepared, she handed out clean handkerchiefs she kept in her hand bag for the boys up front. Maggie uncorked a small brown bottle of nerve tonic, took a dainty sip, and passed the bottle to the front seat.

Sis Johnston was always nearby, with a close ear, when Aunt Maggie and Mother would sit for tea, often for the entire afternoon. Sis would only have to listen quietly to learn the latest news and gossip about Uncle Walter and Bill. "Aunt Maggie, was funny and had a great sense of humor, but she never liked to smile for her pictures, she doesn't look happy in many of her photographs, but she laughed a lot. Uncle Walter liked to drink a bit, from time to time, but I never saw him drunk or anything like that; Uncle Walter enjoyed himself, but did not go to extremes unless he was cussing."

Reaching the west side of town, oogha horns were sounding, from slowly passing Tin Lizzies as neighbors instantly recognized Bill Neal's fancy Chevrolet as it came off the Third Street Bridge into Midland Street's busy commercial district, with the best known captain of the Great Lakes as his passenger. Once the heart of West Bay City, Midland Street was where you could find everything that you were looking for, and perhaps a few quiet places to wet one's whistle as well. Bill honked his horn back as a courtesy to the well-wishing friends and neighbors. The familiar head of thick gray hair capped the tall lanky frame of the Iron Man. The captain smiled, and politely waved his tired hands, to the folks in the street, celebrating his safe, and highly publicized return home.

Curious neighbors who couldn't wait to hear Captain Neal's story jammed the street and stopped traffic for nearly a half an hour before the police showed up. More than just a few men abandoned their cars where they were, in the street, to walk over to Captain Neal's car and see for themselves that he really was still alive. After the beat cops heard the story first hand for themselves, they shook Captain Neal's hand and got everyone back in their cars and traffic moving again. Walter Neal's wife, Maggie, said she was going to get, "the old man" to bed and put a mustard plaster on his chest while she cooked up a Thanksgiving dinner he would never forget.

Maggie was the magician of the kitchen when it came to putting a full spread of vittles on the table. Walter may have taught his bride a few things about cooking for a big hungry working boat crew, but it was Maggie who taught Walter about real home cooking. With plenty to be thankful for this Thanksgiving, Maggie knew it was her prayers that brought her man back home to her.

Only a few heartbreaking days ago, Maggie thought herself to be another widow of Lake Superior, today she was back to her old self, giving the captain orders; after all, it may be the captain's home, but it was her ship. Plucked from the jaws of an icy death, Walter certainly earned himself a dose or two more of Maggie's nerve tonic, before he pulled the thick wool covers over his head for his first real snooze at home since he could remember. Maggie made sure her Walter had enough medicine to ease the pain of his tired, battered and bruised body. There were a handful of other concerned friends and neighbors who also donated to the cause of Captain Neal's recovery with their own bottles, hoping to ease any aches or pains the captain may be suffering as a result of his ordeal.

Worst off were Walter's hands and feet, hammered and bitten by near zero temperatures for 20 hours; he beat the odds and didn't lose any fingers or toes to frostbite. As the last well wishers left the captain's house at 310 S. Erie Street, in Bay City, Maggie was busy in the kitchen; son Bill helped his father get into his pajamas and into bed.

Before the exhausted captain closed his eyes, for a long, hard earned sleep, Bill joined his father with a dose of medicine, and then another, from the half empty bottle. Both Walter and son Bill looked each other in the eye, they quietly studied each others face for a moment, and showed no shame as the two grown men cried together for the second time that day. Both men had looked death in the eye and lived to tell about it. Bill witnessed the horrible deaths of many men in France during the war battling Huns, bullets, mustard gas, and trench foot. Walter, winning a war of his own, beat the Witch of November at her very worst, and lived to tell about it. There was no one from the *Myron's* crew who made it out, they were all dead, all 16 members of his crew, swallowed by the jaws of Superior. Many were baffled, left wondering how on earth Captain Neal had survived and lived to tell of his ordeal when the others had perished. It was the captain himself who knew the answer as to why he survived. "Bill saved my ass; it was my son Bill, who helped me live through hell, and make it back home, said the captain, waving his huge hands to help tell the story," as his niece recalled.

While in France, Bill noticed that many of the military officers were wearing leather trench coats, to fend off the rain, and the cold, of the harsh European winters. It was a resourceful Bill Neal who managed to locate a new leather trench coat and gave it to his father, as a gift, when he got home, from over there. Bill knew, all too well, of the rough weather conditions, his father had to work in. His father, would surely be protected from the elements, as the coat would act as a shield, and protect him, from the piercing cold winds. But little did Bill, or Captain Neal know, that it would be that same leather coat, that would prove to be the difference between life and death. His son's gift, did a fine job, shielding the good captain, from the life stealing cold, saving the veteran mariner's life. The long, well oiled, leather wrap, acted in much the same fashion as a diver's wet suit. Captain Neal believed, the coat had helped trap the precious body heat. Son Bill's gift kept hypothermia from taking Captain Neal's life, as he clung to the remains of the *Myron's* pilot house, as it bobbed in heavy seas.

Captain Neal put his head on the pillow, and quickly fell asleep. Bill stood at the exhausted captain's bedside, for a moment, watching his snoring father. A shadow appeared in the bedroom doorway, it was Maggie, she stood with her hands clenched together in front of her, "C'mon Billy, let him rest, you can help me in the kitchen. We have a lot of work to do before we wake up the old man and put dinner on the table."

There was only one person who could get away with calling Bill, "Billy," that was his mother. After all, Bill was now working as a high level manager at the Chevrolet parts plant in Bay City, Michigan. He was a businessman, a big shot, used to being called Sir or Mister. It was quiet in the kitchen as Maggie and Bill worked together. Still on the table, were the telegrams, reporting the *Myron* going down. Next to it, was the telegram Walter sent from Port Arthur, Ontario, saying he was fine. "Mother, do you think Dad will go back to work on the boats, or do you think, he would take a job at the shop; do you think he would consider it?" I was just wondering the same thing," Maggie said. "We'll talk about it later, now lets get those spuds peeled." At that very moment the snore of an elephant could be heard coming from the captain's bedroom, Maggie and Bill couldn't help it, but they both burst out laughing and crying, at the same time. Dad was home and he was safe. Captain Neal's famous snore, that could rattle every window in the house, came roaring from the bedroom; it was the most wonderful sound the both of them had ever heard. After a few moments had passed, Maggie asked Bill, "Do you think there is a job at the shop your father could do?" Nodding and reassured, Bill said, "Yes, I'm sure we can find the Iron Man something to do over there."

On the other side of town, an enthused Sis Johnston had been collecting and cataloging the local newspapers that made her Uncle Walter front-page news, across the country and Canada, for nearly a week. Sis had a knack for organization and was especially interested in the news coverage of her favorite uncle; she giggled openly with pride when the media dubbed Uncle Walter the Iron Man. Sis and her three brothers had known all along that their Uncle Walter was very special, now the whole world knew it too.

Only 12 years old at the time, Sis had gained an appreciation for the daily newspaper, how its stories were reported and how it kept the whole world abreast of the lost *Myron's* captain, and her doomed crew. A newspaperman had knocked on the door at her house several times; Sis's mother, was Captain Walter Neal's sister, and the newspaper men asked a lot of questions. "Mother said, that those newspaper men were at Aunt Maggie's house too, and they were asking Aunt Maggie and cousin Bill a lot of questions, about Uncle Walter, and his crew. "Sis, and her brothers, found it quite amusing, that they could hear the newspaper men ask, all of those questions, and see them in print a few hours later, the very same day; for the whole world to see. But our hearts were still heavy. Uncle Walter's crew, was still missing, and they were still waiting for official news, one way or the other."

Bay City Times Tribune

Bay City Boat Crew is Missing; Steamer Myron Lost in Gale

Vessel carrying crew of 18, Owned by Blodgett Co. Lies in Lake Superior off Whitefish Point. Captain Neal in Charge of Ship.

So far as known he is only Bay Cityan on Board; Seas drive rescue ship off search for boats. The slender hope held by marine men that some of the crew of the steamer Myron, sank off whitefish point in the gale of Saturday and Sunday might have sur vived, was virtually abandoned at noon today when no further news was received from the various craft searching for a trace of the victims.

Although large quantities of wreckage continue to come ashore this morning no trace has been found of the two lifeboats believed to have offered the only means of escape to the crew of the ill-fated vessel. Reports that men had been seen clinging to the wreckage or lashed to pieces of the Myron's cargo had not been confirmed this afternoon. No word had been received here from the United States submarine 438 which continued this afternoon to sweep Whitefish Bay.

Myron Blodgett of the company owning the wrecked vessel arrived here from Duluth shortly before noon and said he had no verified list of the complete crew of the steamer. He said he believed however that the crew numbered 17. Licensed officers definitely known to have been on board are Captain Walter R. Neal of Bay City, was master, William Lyons of Marine City, Mich., was first mate, R.B. Buchanan of Conneaut, Ohio., chief engineer; Louis Bastian, Tonawanda, N.Y., second engineer and Floyd A. White, address unknown, chief cook.

Went Down in Stiff Gale

Sault Ste. Marie, Mich., Nov 23, 1919; Lake Superior's grave yard has yawned again and today the steamer Myron, owned by the O. W. Blodgett company of Bay City lies on the bottom in four fathoms of water one and a half miles off Whitefish Point while a storm howls from the Northwest. None of the crew of the eighteen men have been heard from since they took to the lifeboats last night as the Myron went down.

For forty five miles up and down the shores of Whitefish Bay from the Crisp point to Vermillion the coast guard are patrolling today in hope some of the lifeboats will be blown ashore. While the beach is covered with lumber from the wrecked steamer, not a body has come ashore at a late hour today. So rough is the weather that it is thought impossible that any steamers have been able to pick up the small boats which have evidently been swept by the storm past Whitefish pint if they have not been swamped in the mountainous sea.

Tow Barge is Saved

The tow barge Miztec which was cut loose from the foundering Myron was picked up this afternoon by an unknown steamer and is now being towed into the shelter at Whitefish point. Her deck load and her rudder are gone but it is thought that her crew is safe.

The Myron was down bound from Munising loaded with lumber. The first word that she was in distress came to the Soo by wire from the Steamer Earling which was lying in shelter and was notified by the Steamer Adriatic which eventually attempted to rescue the crew of the Myron. The Adriatic has no wireless. When the Adriatic locked through here the captain said he had turned around twice trying to pick up the crew of the Myron who were in two lifeboats. The crew seemed unable to hold on to the line thrown to them in the wild sea and as the water was but four fathoms deep the Adriatic had to leave.

May Rescue Some of the crew

Hope that at least some of the members of the crew would be rescued was held out today by marine agents here and captains of steamers who passed this port. Boats that ventured down from the Whitefish shelter yesterday and last night brought reports of men being seen clinging to wreckage or lashed to flotsom from the Myron, although attempts to rescue them in the midst of the enormous seas were futile.

It was believed that some of the crew who thus escaped from the foundering steamer might drift ashore and be revived by the patrolling coast guards, despite their long exposure and terrible buffeting by the great waves.

Word was awaited from the United States sub chaser 438 which spent the night sweeping Whitefish bay and out to the point in search of lifeboats. Tugs and steamers in shelter assisted in the search. Several of the sheltered ships were understood to be equipped with wireless.

Wreckage from the Myron continued this morning to come ashore in great quantity, including the parts of the cabins and the upper works, while the entire bay is strewn with the lumber cargo of the boat and her consort, the Miztec, which was reportedly towed into shelter by an unidentified steamer.

Captain Neal Life-Long Resident of Bay City

Capt. Walter R. Neal, who was in command of the Myron, has been a life-long resident living on the West side. At his home this morning, his son William Neal, said that he did not know of any other Bay Cityans being on the Myron, although it was possible that there might have been others among the crew.

The Myron and the Miztec which was in tow of the
same steamer, wintered here last winter and the Miztec was
taken out in the spring by Capt. Alex McCormick who died
suddenly last spring and who was the brother-in-law of Capt.
Neal.

Capt. Neal, who had lived on the west side from boyhood,
had a wide acquaintance in Bay City. The last letter the
family received from him was written last Tuesday and in it
he had said he expected to be home for Thanksgiving. The
boat had a hard trip up the lake and one which he was then
preparing for was to be the last of the season.

Capt. Neal is survived by his wife, one son, William
and two daughters, Annetta and Irene and four sisters.
Mrs. Alexander Johnston, Anna and Mae of this city and
Viola of Saginaw.

Sis Johnston said she must have read the news of Uncle
Walter's boat and crew about a hundred times, and then, a
hundred times more, wishing all the while it had been a
mistake. The Johnston children, but especially Sis, knew if
anyone could survive zero temperatures, in huge cutting
seas, it would be Uncle Walter.

It wouldn't take but a few hours, after reading the
newspapers, that word of a miracle was unfolding, a few
hundred miles to the north, in Canada, at Fort William. "Aunt
Maggie, was in good spirits when she called, to tell us, she got
a telegram from Uncle Walter, saying he took a long bath, but
was alright. I knew for sure, that telegram came from Uncle
Walter, he always joked about taking a bath, Uncle Walter was
alive! Our prayers had been answered, but could it be possible,
that Uncle Walter, was the only man to survive? Mother said,
we would all have to wait, until the newspaper comes out,
tomorrow to find out the details, but Uncle Walter was alive!"
Captain Walter Neal was in the newspaper headlines, the very
next day. Sis says the newsboys sold out quickly, as friends,
neighbors, and total strangers couldn't wait to hear more about
the missing *Myron*, and her crew.

NEAL PROBABLY SOLE SURVIVOR OF THE MYRON-
Fort William, Ont., Nov 25, 1919

No Trace of Other Members of the Crew is found

Took to lifeboats on order of Captain when water put out
fires under boilers.

Little hope of the safety of the remaining fifteen crew
members of the crew of the steamer Myron which went down
in a gale on Lake Superior, Saturday, is entertained by
Captain Walter R. Neal, master of the lost steamer. Captain
Neal, who was picked up by the freighter W.C. Franz and;
brought to Fort William, believes that he is the solo survivor.

After recovering somewhat from the effects of 20 hours'
exposure clinging to the wreck of the pilot house Captain
Neal told how the Myron, lumber laden and towing the
barge Miztec and encountered the gale early Saturday
morning. Two hours later she was taking water rapidly and
in a sinking condition. When the waters reached the fires
under the boilers the captain ordered the crew, who had
labored vainly at the pumps until that moment, to abandon
ship and take to the yawls.

He saw no more of them for a heavy sea struck the pilot
house and washed it overboard with Captain Neal clinging
to the roof. Two hours later, he declared, a steamer came close
to the wreckage, hailed him and promised to send a boat but
passed on. Though with little hope of life, he clung to the
pilot house in the heavy seas for hours, drifting about in a
semi-conscious condition until the Franz rescued him.

NOT ON PARISIAN ISLANDS
Sault Ste. Marie, Mich., Nov 26, 1919

Search there futile, Crew is Probably All Lost. Practically
the last hope of the owners of the shipwrecked Myron as a

sheltering place for possible survivors of the crew, was combed
by coastguards Tuesday without a trace of the 16 members of
the crew who took to the lifeboats and clung to the wreckage
as the ship went down about 10 miles off Whitefish Point
early Saturday evening.

"Good news, about Uncle Walter, was worth the wait, and Mother was so happy, she was singing, but fretting again, over getting Thanksgiving dinner ready on time. There was certainly plenty to be thankful for this Thanksgiving. Maggie, rang Mother up on the telephone, she said that she was going to take Billy on the train to meet Uncle Walter at the Soo. But the news was still bittersweet; what about the families of the other men who were still lost, what kind of Thanksgiving will they have?"

There was little more than a few lines about the tragedy in the next day's papers, but more tragic news was to come into print. Aunt Maggie and Cousin Billy were already on the train heading for the Soo with a sack full of food and a little liquid refreshment for Uncle Walter. I figured if I kept praying, like I did for Uncle Walter, maybe I could help bring those missing men back," said Sis.

News Chronicle / Port Arthur
HOPE IS ABANDONED FOR MYRON'S CREW
Sault Ste. Marie, Mich., Nov 27, 1919

With a terrible northeaster accompanied by a blinding
snow raging over Lake Superior, the search for the sixteen
members of the crew of the steamer Myron, which sank near
White Fish Point, Saturday was given up Wednesday, the
searchers unable to continue in the gale and convinced that
members of the ill fated crew have been drowned.

The blizzard which Tuesday night, caused government
coast guards to abandon the search. It had previously been
hoped that they might still be afloat in the non-sinkable

lifeboat from the Myron. Today's searchers declared that the crew could not possibly have survived the storm last night. Three lake vessels are reported grounded at Pipe Island shoal, White Fish Point and Cedar Point reef and it is expected that shipping losses will be heavy.

"It was the only news clip, I glued to that page of my scrapbook. I made sure, I left plenty of room, just in case we heard some good news later, that his crew was rescued. But there was no good news, for the lost crew, their bodies were coming to shore, it was simply too late, for them."

"It was a wonderful Thanksgiving, the best ever, Uncle Walter was safe, and we had everything to be thankful for. Uncle Walter called the house, to thank me again, for all of the prayers I said for him. He even told me that he loved me and that I was his favorite niece." I told him, "I love you too, Uncle Walter, and I'm glad you're back home." Aunt Maggie told Mother, that Uncle Walter ate like a horse, during their Thanksgiving dinner. But between bites, Uncle Walter cursed the captain, that would not throw him a line." Sis said, "We all lost count, of how many times, Uncle Walter called the captain of the *McIntosh*, a 'son of a bitch.' We, were all quite happy, to listen to the captain's cussing, and have him tell us his story, over and over and over again. We had all heard, Uncle Walter, tell the story so many times, we could all tell the story, word for precious word," his niece said with a giggle.

Chapter Eight

After The Storm

Just before noon on Saturday, November 29, 1919, a battered, bruised, and very thankful Walter Neal slowly swung his long frame out of bed, to put his stocking feet on the floor. His prayers, had been answered, and he was, by the grace of God, safely back home, in his own bed, and with his family.

Sore in places he had never been sore before, the joints in his feet cracked, as he took his first steps of the morning, across the linoleum floor. Maggie Neal had been working quietly in the kitchen, while "the old man" slept. As he made his way to the kitchen, Walter heard the sound of heaven, as the bacon in Maggie's fry pan sizzled, on the stove. Normally, when Walter came home from his boat, Maggie, would insist that the crusty captain take a bath, before coming into her kitchen. This time, Maggie didn't say a word. "Aunt Maggie, told Mother, she laughed so hard, she cried, after Uncle Walter, told her, that he wanted to get into a hot tub, and soak, before he ate his breakfast," Sis said with a grin.

Maggie knew her husband's escape from death was a miracle. But volunteering for his own bath, were the last words, she thought she would ever hear come out of his mouth. Aunt Maggie set a cup of black coffee on the kitchen table in front of the chair closest to the stove so Uncle Walter could stay warm. The kitchen counter tops were covered with all types of food brought over to the captain's house, the day before. Their many friends and neighbors tried to help the Neals celebrate a late, but wonderful Thanksgiving.

The kitchen table was empty except for Captain Neal's elbows, his cup of coffee, and two very small but wonderful pieces of paper. One was the dreaded telegram that brought tears to Maggie Neal's eyes with the message she had prayed so hard not to receive. The other was the return telegram that brought a wide smile to Captain Neal's wife's tear stained face. Maggie said she "had to reread the telegram from Uncle Walter several times before the words sunk in; she kept saying it was a miracle."

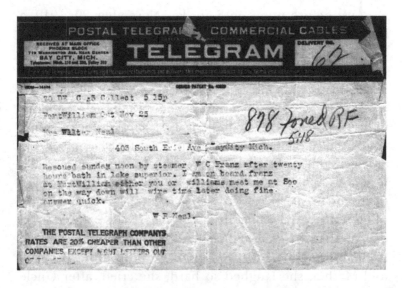

Telegrams / Courtesy Great Lakes Shipwreck Historical Society

Telegram #1

Fort William Ont. Nov 25

> *Mrs. Walter Neal 403 South Erie Ave. Bay City, Mich.*
> *Rescued Sunday noon by steamer W.C. Franz after*
> *twenty hours bath in lake superior. I am on board franz at*

*FortWilliam either you or William meet me at Soo on the way
down will wire time later doing fine answer quick. W.R.
Neal*

CANADIAN PACIFIC R'Y. CO.'S TELEGRAPH
TELEGRAM
CABLE CONNECTIONS TO ALL PARTS OF THE.WORLD
J. McMILLAN, Manager Telegraph, Montreal.

83 WN W :25

BAY CITY MICH NOV 24TH 1919   112

CAPT WALTER R NEAL.

STR W.C.FRANZ. FORTWILLIAM

OUR PRAYERS HAVE BEEN ANSWERED   MOTHER AND I WILL MEET

YOU AT THE SOO ON   WAY DOWN LET US KNOW

WHEN YOU WILL BE THERE.

W.J.NEAL.

1905

Telegrams / Courtesy Great Lakes Shipwreck Historical Society

Telegram #2

From: BAY CITY MICH NOV 24TH 1919

CAPT WALTER R NEAL.

> *STR W.C.FRANZ FORTWILLIAM*
> *OUR PRAYERS HAVE BEEN ANSWERED MOTHER*
> *AND I WILL MEET YOU AT THE SOO ON WAY DOWN*
> *LET US KNOW WHEN YOU WILL BE THERE.*
> *W.J.NEAL*

As Maggie poured "the old man's" second cup of coffee,
Uncle Walter was on the phone with his sister Nettie, at the
Johnston house. "When he finished speaking with mother I
got to talk to him on the phone for a moment," Sis Johnston

recalls. "Uncle Walter said he would show me the bruises on his legs if I wanted, but he was my hero and I didn't want to see him hurt anywhere. Heroes don't get all black and blue, I didn't want to look at that. I was just so happy Uncle Walter came back home. He was very, very special to my brothers and I; he was a lot of fun, and we just couldn't imagine what it would have been like without him. We cried when we got the news of his boat going down, but somehow we knew that Uncle Walter would find a way to get back home; we knew he was tough and he would make it back to dry land."

Friday's Bay City Times was still on the arm of Captain Neal's over stuffed easy chair in the living room; it hadn't been opened yet. Maggie brought the newspaper back to the kitchen and handed it to her husband who had downed his second cup of coffee and was ready for another. On the front page of the "Times" was the story about Captain Neal's return home that the newspaperman had written after meeting Uncle Walter's train at the Bay City station. "Uncle Walter may have nearly lost his life but he never lost his sense of humor. Not shy in the least, Uncle Walter enjoyed reading about himself in the papers, and would often read the articles out loud for us."

The Bay City Times
November 28, 1919

Sole Survivor of the Myron Reaches Home

    Capt. Neal Tells of His Long Fight for Life in Lake Superior
    Was safe for hours before the News of the Wreck of His Vessel Reached the public.

    Capt. Walter R. Neal, sole survivor of the officers and crew of the steamer Myron, wrecked in Lake Superior last Saturday, arrived home today, accompanied by his wife and son, William who met him at the Soo.

Capt. Neal is suffering comparatively little from the effects of his 20-hour exposure in the water buffeted about by a storm, and he insists that within a few days he will have entirely recovered.

"I never believed that a man could stand so much exposure and come out of it alive," he said this morning. "They tell me that I have the record for all Lake Superior on that score, and if I have I am perfectly willing to let it stand without any attempt to make a greater one. They called me the Iron Man up at Fort William, and of course I suffered intensely from the cold. I was either in the water up to my knees or it was washing over me for hours and my feet and legs were swollen terribly but I never even caught a cold. But how good that steam radiator on the Franz felt to me. I just clung to that for hours after they took me aboard without even taking my clothes off. "If the Franz, which picked me up, had been equipped with wireless the news of my rescue would have reached home even before the news of the wreck was sent out, for I was half way across Lake Superior on the way to Fort William when you got the news of the wreck here"

"The captain of the Franz and the crew treated me fine. They couldn't have done more for any man, and after he had rescued me, the Captain cruised about for some time in the hopes of finding some other survivor, but without success. I had drifted miles from the wreck when they found me, and was so close to the lighthouse on Parisian Island that I could distinguish the windows in the dwelling, there is seemed just a short distance as I looked at it just as if they ought to have been able to see me, but I found out afterword I was four miles off shore and of course they couldn't distinguish a man at that distance."

"Uncle Walter would get quiet every once in a while when he talked about his crew from the *Myron*; sometimes under his breath you could hear him say, 'May god bless their souls.' But he usually said that when he didn't think anyone could hear him; we always prayed for those fellows on Sunday," Sis said.

Now in his 50's, Captain Walter Neal had the entire winter of 1919-1920 to decide if he wanted to tempt fate and find a boat that needed a skipper, or take his son Bill's suggestion and go to work at the Bay City Chevy plant on the shore of the Saginaw River. And for Uncle Walter, in view of the situation he had just put behind him, working at the Chevy plant sounded like a "damn good" idea, at least for now anyway.

The Iron Man was right, he did get better as the days went by and he was getting back to his old self. Sis said, "Uncle Walter would wink at me when he walked over to the bookshelf behind his chair and took a nip from the little bottle he kept hidden behind the books on the top shelf in the living room." The days and weeks seemed to fly by and it was almost springtime, most of the snow had melted and the ice on the river was beginning to break up. With the memory of the loss of his crew, his boat and of the grim reaper's blade of death coming so close to him in icy Lake Superior, Captain Neal took the job at the General Motors parts plant that his son Bill managed. "Uncle Walter said he did the best he could to work there, but working inside was driving him nuts," Sis said. "He said he made good money, and his job wasn't too hard, but he just couldn't stay inside all of the time." It got even tougher for the leathery old sailor to just sit and watch when the boats were underway again in the spring. "Uncle Walter said he had to get back on a boat and stay on a boat until he was an old man, and he was already starting to get up there in years. Uncle Walter wasn't a quitter so he stayed on the job at the plant and tried to make a go of it for as long as he could, but there were some problems. He would walk to work everyday across the Third Street Bridge to get to the manufacturing plant."

When the river ice began to break up and the boats started to move, Uncle Walter, could not help himself, he would stop on the center section of the bridge and watch the boats head out to the Bay. Uncle Walter liked to ride the section of the bridge that swung open, to allow the working steamers up and down the river. Uncle Walter simply couldn't walk away, he would often end up reaching into his lunch pail for a sandwich and pour a cup from his jug as he watched the boats come and go from his perch at the bridge rail.

After a little more than too much daydreaming, the captain would slowly regain focus on his task at hand and walk the last few blocks to work. Unfortunately for son Bill, number two man at the plant, he couldn't tolerate his father's sustained tardiness much longer and be fair to the other men in the shop who were already grumbling about the boss's father. Consequently after only a few short weeks as a shop worker for General Motors, Walter Neal, Iron Man of the Great Lakes, the benefactor of a miracle from God, the sole survivor of the shipwrecked *Myron*, the man who was able to cling to the wreckage of his ship for 20 bone-gnawing hours, was fired, by his son, for coming in late every day for the last several weeks.

The 1920 Great Lakes shipping season had begun and every working steamer that would be in service that season already had a skipper. Walter Neal managed to land a job with his former employer O.W. Blodgett as a first mate on the *Zillah*, another wooden, time worn workhorse that still had plenty of life left in her. But for Walter Neal, even with a new assignment he would never forget the faces of his 16 lost crewmen; or would he ever forget the *Myron's* last desperate moments, or the 20 endless freezing hours he thought might have been his last.

Maritime sketch artist Ed Pusick, with pencil and paper, and his hand guided by the spirit of those lost souls, envisions the *Myron's* last moments, before she finally gave in to an angry Superior, and plunged to her death.

The Myron's last moment, sketch by Ed Pusick /
Courtesy Great Lakes Shipwreck Historical Society

"Captain Walter Neal said, 'he had no problem working as a first mate aboard the *Zillah*,' he was back home on the lakes, where, he belonged," says niece Sis Johnston. Things eventually did get back to normal for Walter Neal, but for Uncle Walter, normal was a relative term. In retrospect, turmoil and stormy weather, land or sea, is where Walter Neal felt most at home, or at least that's where he could be found a good part of the time. There would also be several more homecomings that would come to pass before Uncle Walter was finally retired from his service as a merchant seaman.

Uncle Walter often shared his stories with those who would inquire and many times for those who hadn't. "Uncle Walter never stopped talking about the *Myron*, or that 'son of a bitch' from the *MacIntosh* who wouldn't throw a shipwrecked man a line."

It wouldn't take long for fate and fiercely respected Lake Superior to again touch the very soul of the Iron Man. In May of 1921 the schooner barge *Miztec*, after escaping death behind the *Myron* in 1919, would now have her own date with destiny, and more men would die. Walter Neal was again in harms way, in a storm on Superior, a few miles off Whitefish Point, once again. What was about to happen next is the kind of stuff that legends and movies are made from, stories that could never happen, but seem to happen over and over and over. On May 14, 1921, newspapers across the Great Lakes carried news back home to the sailors' families that Superior had swallowed more lives, in just one gulp. Maggie Neal and the Johnston's would again learn that Walter had gone through yet another storm. "The old man got in some trouble again, but he was O.K.; it seems like we heard Aunt Maggie say that a lot when we were kids," said Sis. "Walter was O.K. and there was no need to send a telegram saying he was fine; Maggie said telegrams are for delivering bad news and she's already had her share. Uncle Walter always said Maggie hated getting telegrams and it's not terribly difficult to understand why."

It was a sunny morning in mid-May, and Michigan's spring air still carried a chill. Maggie Neal, with her sweater worn over her shoulders like a cape, stooped down to pick up the newspaper from the front porch; there on the front page of the Bay City Times, she would learn that "the old man" had another tough day at work.

The Bay City Times
May 14, 1921

Barge Miztec Wrecked Near White Fish Pt.

Miztec and Peshtigo Break Away From Steamer Zillah;
Latter in Shelter; Peshtigo Riding at Anchor in Storm.

Sault Ste. Marie, Mich., May 14
Six men comprising the crew of the barge Miztec are
believed t have been lost in the storm of last night and early
today which tore the Miztec and the Barge Peshtigo loose
from the steamer Zillah off Whitefish Point in Lake Superior.

A wireless report from Whitefish Point this afternoon said the Miztec was wrecked, but the Peshtigo was riding a heavy sea at anchor a mile off the Whitefish shore. This report was confirmed by the steamer Renown, which locked down this afternoon. The Renown reported passing the wreckage of the Miztec three miles off Whitefish this morning. The steamer turned about to go to the rescue of a man seen lying on the roof of a cabin, but as the big vessel approached the man rolled off the wreckage into the lake. No other bodies have been sighted.

Earlier advices said the Zillah, of the Blodgett line had found shelter in the lee of Whitefish Point.

Fear for the Peshtigo

The Miztec was in command of Captain K. Pederson, of Buffalo and the Peshtigo in charge of Capt. Dan Campbell of Tonawanda. A heavy wind was still driving the snow in a fierce blizzard over the lake at noon today, and the fears are expressed here that the Peshtigo may not be able to weather the storm. The Peshtigo like the Miztec, is of wooden construction and carries a crew of six men.

It was the barge Miztec that was in tow of the ill-fated steamer Myron of the Blodgett line, when that vessel was lost off Whitefish Point on Nov. 22, 1919, with the loss of 16 lives.

The Zillah is reported safe but the barges have not been located. A Coast Guard cutter has been sent out to search for the barges. No other mishaps have been reported as the result of the blizzard, one of the worst in years. Six inches of snow fell during the night and the temperature dropped to below freezing.

"I haven't heard anything more than is given in that dispatch," said Myron Blodgett, one of the owners, this afternoon. "I am confident however, that the barges will ride out the storm, as there is good anchorage in that vicinity."

Capt. Walter Neal, of this city, is sailing as mate on the Zillah.

Maggie Neal took a deep breath, and slowly let it out as she raised her face to the morning sun. "Well, at least 'the old man' is still alive," said Aunt Maggie, "life was back to normal at the Neal house." Aboard the *Zillah*, Uncle Walter was counting his blessings. *The Myron*, just a year and a half ago, went down some 3 miles west of Whitefish Point. Now in almost the same spot, he was witnessing the death of the *Miztec* as she shuddered before heading for Superior's sandy bottom taking the lives of seven crewmen with her.

In the coming shipping seasons, Walter Neal would regain his position as captain, and live a relatively quiet but busy life for the next decade or so, but there was another storm brewing on the horizon for the captain. This time, the captain would be weathering a legal whirlwind in the wake of his own arrogance, in a federal court of law. While Walter Neal reveled in the instant respect and admiration he received from those familiar with the story of the *Myron's* sinking, Neal also had the savvy and wherewithal to make his boss a good deal of money each season. And that is where, and how, Uncle Walter found himself in trouble again.

The Detroit News
August 1935

Capt. Neal, Lake Veteran, Weathers a Legal Storm

> *Man and boy, Captain Walter R. Neal, of Bay City, has sailed the inland seas for 52 years and has been a Great Lakes skipper for 41. Tuesday afternoon, aged 70, the captain jauntily left the Federal Building, his walk seeming to indicate that he was at peace with the world.*
>
> *A little earlier in the day it looked rather serious for Capt. Neal, who stood before Capt. William T. Kiel and William H. Dungan, inspectors for the United States Steamboat Inspection Service of the Department of Commerce. Capt. Neal was faced with a suspension of his master's license after two score years without a mark against his record.*

The charges on which the ancient mariner was tried
were brought by Maj. R.C. Crawford, the United States
district engineer of the War Department here. They involved
a happening in the Detroit River the afternoon of June 25.
The scene was Ballard's Reef Channel, opposing Grosse Ile, a
section of the river being deepened by a dredge.

3 SHIPS ABREAST

Capt. Neal that day, in command of the Liberty, light
and bound for Cleveland, overtook and passed the William
B. Schiller, also down bound. Maj. Crawford charged, just
as another ship, the J. Oswald Boyd, passed, upbound. The
situation presented was three steamboats abreast in the
Ballard's Reef Channel, in coatravention of the Rivers and
Harbors Act.

Maj. Crawford also charged:
Instead of checking his speed to five miles an hour, he
increased it. His ship tore down some of the channel range
lights. He failed to heed danger signals. A parade of witnesses
came to the stand. Finally Ernest W. Johns, inspector on a
Government drill boat, took the stand. He was an eye witness.
Kiel questioned him. Didn't Capt. Neal, he asked, go around
the channel marker and cut out of the channel?

"He did", said Johns. "Well", said Frank X.
Norris, assistant district attorney, "If I am not
mistaken, Capt. Kiel, a boat on the outside of the
channel is not governed by the harbors act". "That's
right", said Kiel.

NORRIS WALKS OUT

"I guess then", said Norris, "there's nothing more
for me to do around here".

Norris was on hand because, if the captain was
convicted, he and the employer, the Nicholson Transit Line,

would be liable for charges before the United States District Court.

Capt. Neal took the stand, emphasizing he left channel and saying that he was able to do so because his boat was not loaded. He did so, he said, because another boat, the Ajax, was giving forth so much smoke as to darken the scene.

The court adjourned to study the record. A decision will be forthcoming within a few days.

Capt. Neal didn't have a chance to tell the inspectors the story of another great crisis in his marine career, about the day back in 1919 when he was the skipper of the Myron and got caught in a gale off Whitefish Point. He sent his crew of 16 off in the boats and then decided to go down with his ship. The next day he was found by a rescue boat, unconscious and floating on cold Superior's waves in part of the Myron's pilothouse, where he had planned to meet his death. His 16 sailors were lost.

It took several days for Captain Walter Neal's case to be reviewed and decided by the United States Steamboat Inspection Service of the Department of Commerce. At the age of 70, Captain Neal was still going strong and had a number of good seasons left in him. But if the Steamboat Service decided that the skipper was in violation of the law for his actions, his license could be taken away and he would be forced to either retire or find another means of making a living. Cocky, confident and with a bit of a swagger in his step, Uncle Walter seemed to already know the outcome. Just a few days after testimony from eyewitnesses was taken in that federal courtroom in Detroit, the judges had come to a decision. Captain Neal, his employer and his attorney were summoned to court to hear the government's verdict. If found guilty, Captain Walter Neal, Iron Man of the Great Lakes, might find his master's license revoked. Could it be possible that one of the most familiar, storm-worn captains on the Great Lakes might never command another vessel?

Or would there be another last moment miracle that would arrive in the nick of time to save Uncle Walter's hide and have him come up from the slop bucket, smelling like a rose?

The Liberty / photo courtesy Ralph Roberts

The Detroit News
August 17, 1935

Veteran Skipper Gets Clean Slate

> *Captain Neal, 70, is Freed of Charges*
> *A forty-one-year record as a Great Lakes skipper remained unblemished Friday when steamboat inspectors reported that charges preferred against Capt. Walter R. Neal, 70-years old, were not sustained.*
> *Capt. Neal was brought before Inspectors William H. Kiel and William Dungan, on charges preferred by Maj. R.C. Crawford, United States district engineer here, of piloting the steamer Liberty through the Ballard's Reef on the Detroit River in violation of the Rivers and Harbors Act.*

He was charged with going more than five miles an hour, overtaking another boat, coming abreast of two others in the channel, disregarding warning signals and running down some range light buoys.

Testimony disclosed that the Liberty was out of the channel where dredging was in progress and, therefore, was not subject to regulations of the Rivers and Harbors Act.

The transcript of the testimony will be sent to Fred J. Meno, superintendent of the steamboat inspection, for final approval.

"After breaking nearly every rule in the book, and getting away with it, Captain Walter Neal truly was the 'Iron Man.' A captain, by any other name, would have likely sailed his last ship. But Uncle Walter always had good luck on his side," recalls his "little Sis." But by this time, "little Sis" was a fullgrown woman, she was teaching high school and had her own car. Sis Johnston said, "Uncle Walter behaved himself for the most part until he retired. He was nearly eighty years old when they retired him from his command, and with a good pension. Uncle Walter would have stayed on board until they carried him off on one," Sis said with a firm and convincing nod of her head.

"Father thought the steam boat inspectors should have nailed Uncle Walter's rear end to the wall, after he knocked out all of those range lights while at the same time, breaking every rule in the book, and he wasn't happy at all about the final decision to dismiss all charges." Sis recalled that as her father and Uncle Walter got older they didn't swear as much at, or about each other. Their once seemingly simple Victorian world was fading and the world was becoming a faster moving and more dangerous place.

But for Captain Walter Neal, life was good; he was enjoying the fruits of being a grandfather, and took great pride in watching his grandson and granddaughter grow up.

Captain Walter Neal with grandson James,
note rail tracks on deck / photo courtesy
Great Lakes Lore Museum

Buffalo New York Newspaper

NOTED SKIPPER SUPERVISES JOB ON CRAFT HERE

Captain Walter Neal will command altered Collier in
Buffalo-Montreal grain trade.

Capt. Walter Neal, veteran steam boat skipper and
hero of many exciting adventures on the Great Lakes, is here

looking after alterations upon the steamer Marquette and Bessemer No. 1 at the Buffalo drydock. She was built in 1904 at the same plant as a collier, equipped with the tracks on her deck on which coal cars were shunted aboard to dump their loads into the hold through the hatches. She has been carrying coal on the lakes ever since, but hereafter will be engaged in the grain trade between here and Montreal. It is expected she will get away with her first cargo some time today with Capt. Neal in command, W. E. Fitzgerald representative of the company said yesterday. The steamer has had the tracks and all other superfluous equipment removed from her deck, and the apron taken from the stern.

Capt. Neal, who is one of the best known skippers on the Great Lakes has had a career that has been replete with thrills. His most outstanding adventure, probably, was when the little lumber steamer Myron was lost on Lake Superior, near Whitefish Point in November, 1919, with the entire crew, excepting Capt. Neal, who left the ship last and managed to cling to the pilot house for more than twenty hours as it was buffeted by heavy seas in the frigid waters to Lake Superior. He managed to keep alive by moving about on the piece of wreckage. His hands became encased in ice, but he managed to cling, although the pilot house was tipped over several times, and the doughty skipper was finally picked up in a semi-conscious condition by the steamer W. C. Franz, commanded by Capt. Jordan, the same man who last December was wrecked on the steamer Agawa in Georgian Bay. He and his crew were rescued by tugs and fishermen after several hours in the ice covered cabin.

Capt. Neal commanded the former Buffalo steamer Arizona when she was one of the few ships that rode out the tremendous gale of 1905 when several big steamers were lost on the lakes. He is in the best of health and spirits at present and being a man of exceptional ruggedness he expects to sail for several more years.

The remaining years of the 1930's were clouded with front-page war news from Europe and Africa. Italian dictator Benito Mussolini had sent his troops to capture Ethiopia in the northeast section of the African continent. Nazi Germany's Adolf Hitler aggressively launched the Second World War on September the first of 1939, invading Poland. The world as most sailors had known it was fading fast; the once huge fleet of wooden steamers that worked the Great Lakes was disappearing. Some died from old age and still lie on the bottom where they fell far below the surface, some are little more than a few planks and a rusting boiler; others that remain identifiable seem to wait in their silent watery haunts for her crew to make repairs and get her back to work. Many of the other once proud workhorses were cut down into barges, scuttled or abandoned and burned to the waterline in some out of the way place and forgotten. Many maritime historians say, that if you take your time and look close enough, and ask a few questions, you can rediscover the remains of countless old steamers all the way from Cornwall to Buffalo, Ashtabula to Chicago and Thunder Bay. These were now modern times; the best ships are being made of steel and every vessel was now required by maritime law to have a working wireless, with a man at the key. Many old sea veterans still weren't convinced that steel was the best material from which to build ships, after all, wood floats and steel sinks. And there were still a good number of other skippers who thought having a radio on board was somewhat of an insult to their skill and expertise in guiding their vessels. After all, in the minds of some, they had done quite well over the years with their watch, compass, and dead reckoning, and without the aid of a radio onboard for their entire career. Soon to be gone forever were the hand built steamers of creaking white oak, that once relied on firing colored flares into the air to send a message past her rail, past the sound of the captain's megaphone.

An aging Captain Walter Neal managed to change with the times, and remained the tough, leathery sailor he had always been, just a little slower perhaps. In January of 1941 Captain Alexander Johnston, Walter Neal's brother-in-law died, "Uncle Walter was there for me when Father passed away. Uncle Walter had been like a second father to me, he would always call and see if I was all right, or if I needed help with anything; he was always very special. I remember him saying a prayer out loud at Father's grave side, and he meant it, even though the two of them bumped heads over nearly everything two men could disagree about for most of their lives. Aunt Maggie and Uncle Walter came to the Masonic Hall for a dinner after Father's funeral. She said "the old man" was going to put in one more year and then retire."

When the 1941 season opened up on the Great Lakes, everyone knew that America would soon enter the raging war in Europe. Many Americans wanted to remain neutral and stay out of the war, even though they were supplying the British with ships, planes and bullets. American industry was converted to wartime production, the steel mills needed ore and coal, and the factories needed materials for tanks, bombers and guns, and it would be the skippers on the Great Lakes who would keep the materials moving and the mills and factories working.

Uncle Walter said, "without my boat, no one would be working, I was hauling the coal that fired the machines that did the work," and he was right. Shortly before the close of the 1941 season, Uncle Walter wanted to sign on for another year of service after the December 7th Sunday morning surprise attack by the Japanese at Pearl Harbor. Captain Neal's son Bill had fought the Huns during the first war, and Uncle Walter said he had plenty of fight left in him to do some damage to the Nazi Huns now. But Captain Neal had served his last season on the Great Lakes, it was time for the "Iron Man" to step aside and make way for a younger man.

The Bay City Times
April 29, 1942

> As his command of last year, the steamer Fleetwood
> prepared to leave Ecorse today, Captain Walter R. Neal is at
> Bay City, Mich. helping his wife with the spring cleaning.
>   The 77-year old master is off the lakes after 60 years of
> sailing. The lone survivor of the steamer Myron that went
> down in November, 1919, off Whitefish Point, Lake Superior,
> Neal clung to the remnants of the pilot house for 20 hours
> before being rescued by the steamer W.C. Franz. But he faced
> his greatest Lake Superior storm November 28, 1905, as
> master of the small steamer Arizona, the only vessel to enter
> Duluth harbor undamaged during the gale that broke up
> the Maraafa just outside the harbor.
>   Captain Neal went with the Nicholson Transit Co. 12
> years ago as master of the Liberty, transferring to the Fleetwood
> in 1940.

Maggie (McCormick) Neal, would enjoy the company
of her crusty old man Walter, for the next seven years, winter
and summer. Sis Johnston has repeatedly said that, "Aunt
Maggie had a really good sense of humor, she just hated to
smile for the camera. Uncle Walter kept himself busy, one
way or another, but he liked to use his jackknife and whittle
out little things with his hands." It also came to light that it
was Uncle Walter who taught Sis's brother, Chuck, how to
use a jackknife to perfectly carve his initials into a piece of
wood, as an engraver would scribe a piece of gold. And, it
also appears that "Uncle Walter" was the one who helped
his nephew, Chuck, cut and shape the little star wheeled
noise maker that got his nephew into "hot water" with
mother, and the neighbors on occasion.

Sis said, "As he got older, Uncle Walter loved to share his
stories about working the boats. You couldn't sit with him
for more than a few moments when he would bring up the
Myron, and that 'son of a bitch' who left him to die. Uncle

Walter would say, 'I showed that son of a bitch, I showed him, they took his damn license away and made it law to pick up survivors, that son of a bitch.' Uncle Walter had a certain way of speaking, he would curl his words at the end, so every time he practiced the expert command of his adverbs and adjectives, it almost seemed comical. And while we never swore like Uncle Walter, he always brought a smile to our faces when he was focused on a topic that did not bring him favor."

In their retired years together, Sis recalled, "Aunt Maggie and Uncle Walter would go out to the Casino at Wenona Beach, where they could have a little fun and enjoy worldclass entertainers. Being proud grandparents, they would occasionally take their grandchildren to the amusement park for a ride on the jackrabbit roller coaster or on the electric bumper cars. Wenona Beach was a Victorian classic, its post cards, sold in the gift shop to tourists, depicts the fully clothed women and the cloth covered chests of men enjoying the refreshing water and sandy beach. The amusement park was often a daylong event, to allow for a swim, a ride on every one of the mechanized amusements; and when it was time for a break, you had lunch. Everyone took a short nap in the shade under the huge willows that lined the shore near the amusement park. After our nap, we got back in the water for an hour or two until it was time to go home, we were tired." Wenona Beach is just a memory in the 21st Century, but whose image, in its heyday, has been captured in photographs. Wenona Beach was located just west of the mouth of the busy Saginaw River.

In 1949, life for retired sea skipper Walter Neal changed forever. The mother of his children, the hater of telegrams, and his best friend Maggie, passed away.

Mrs. Margaret A. Neal was eighty five years old when she became ill and died at the end of October in 1949. Maggie was born in Amherst Island, Canada, on October 2, 1864; she married Walter in Bay City, Michigan, on December 21, 1892. Maggie and Walter would have celebrated their 57th wedding anniversary in 1949. Maggie was the best life's mate

"Uncle Walter" could have had. Maggie was also from a sailing family, and knew the rigors of a maritime family's life. Maggie's father, husband, brother, and nephew were all sailors; it's what she knew. For Walter Neal, Maggie was all he knew for the last seven years. He spent his entire life at sea, and a few cold weeks each winter at home. Uncle Walter had brushed the hem of the grim reaper's hooded robe more than once and lived to tell about it. But his more exciting days of escaping death and the law were behind him; he was now retired and devoting a great deal of his time to his bride Maggie. And his bride was now gone forever.

Sis said, "after Maggie's passing, Uncle Walter would often say, 'she's taking care of my boys (from the *Myron*) now, and it won't be long before I can give them their orders, personally.' Somehow in the back of our minds we knew that Uncle Walter, when it was his time, would probably find a way to get in trouble 'up there' too! Just over a year after Maggie passed away, it was Uncle Walter's turn. No one will certainly ever have a better attended funeral, than Uncle Walter!"

The Bay City Times
Wednesday March 21, 1951

Veteran Bay City Lakes Skipper, Captain Walter R. Neal,
Dies at 86

Capt. Walter R. Neal, longtime Bay Cityan who sailed the Great Lakes for 61 years, died Tuesday evening at General Hospital at the age of 86.

He resided at 310 South Erie avenue and had been ill for the past two and a half months.

Coming to Bay City with his family at the age of 16 from Chatam, Ontario., where he was born March 1, 1865, he began his sailing career a year later as a second cook aboard the Steamer Nelson Mills, of the Mills Transportation Co.

For 10 Years, he continued to serve aboard that ship as it carried lumber and lath to Tonawanda, N.Y., from the *old Folsom and Arnold mill at the Southeast approach of the Belinda street bridge. During that time, he rose to the rank of first mate.*

Capt. Neal's first command was aboard the steamer George King. Subsequently, he was skipper of the Arizona on which he weathered what he considered the worst storm of his career on Lake Superior in 1905; the Cury, the Sacramento, of the Davidson fleet and the steamer Myron owned by Capt. O.W. Blodgett, of Bay City.

It was in November of 1919, that Capt. Neal became the only survivor of the Myron when she went down near Whitefish Point with 16 seamen aboard. The captain survived by remaining afloat for 20 hours in near-zero temperatures on remnants of the ship's pilot house. He was rescued by crewmen of the Steamer W. C. Franz,

After the Myron sinking, the veteran seaman took command of the Steamer Marquette and Bessemer No. 1, operated by the Lake Erie Coal Co., before becoming captain of the Nicholson Transportation Co's Motorship Liberty.

He continued in command of the Liberty for 10 years, being assigned to his last command aboard the company's Fleetwood in 1940. Since his retirement 10 years ago, Capt. Neal had busied himself with the building of model ships, among them a miniature of the S.S. South American, a popular cruise ship now operating on the Great Lakes.

The captain was a member of the Grace Episcopal church and a life member of Wenona lodge 256, F. and A.M.

His son William and daughter Irene survive him. Captain Walter R. Neal was buried, at Maggie's side, near the chapel at Elm Lawn Cemetery on Ridge Road in Bay County, Mi.

Visiting Elm Lawn Cemetery is like taking a walk back in time. Sis would say she always enjoyed the visiting the family plot, to plant flowers and trim the shrubs; and she liked to look at the stones and mausoleum's that were designed to last for eternity. The Johnstons and the Neals and their immediate offspring rest at Elm Lawn. Numerous remnants of Captain Walter Neal's legacy can still be found, from Whitefish Point to his home town and elsewhere; and most are available for public viewing. The huge white oak rudder salvaged from the steamer *Sacramento* stands in tribute to the men who built her near the shore of the Saginaw River. The rudder is mounted in its proper upright position, and is located in the heart of the former grounds of the Davidson Shipyard, on the west bank of the river. Battling time and the elements are the bones of a half dozen wooden hulled merchant vessels that lie rotting and helpless in the shallow water near shore.

The Sacramento's oak rudder / photo Richard VanNostrand

Rotting hulls during low water on the Saginaw River /
photo Eric Jylha

For those persons who would like a closer encounter with Captain Walter Neal and his most famous command, the *Myron*, set sail for the Great Lakes Shipwreck Historical Society's Museum located at Whitefish Point, Michigan, on the northern shore of the state's Upper Peninsula. There you will find the ship's brass inspection plate and her wheel; you will also discover the *Myron's* whistle, her glasses and a piece of china used by her crew. The *Myron's* anchor lies in the hold of the *Valley Camp* Ship Museum at Sault Ste. Marie, Michigan. The rest of the *Myron's* remains still lie below the surface of Superior, several miles off Whitefish Point, accessible to scuba divers, but protected by law from souvenir hunters.

Chapter Nine

The Chair, the Charm, and the China Doll

Could it be that this writer was the luckiest, of the thousands of Alexandra Johnston's students? Perhaps I was, maybe I am. The "Captain's Chair" project began as an effort to write a one page history of the house I had hoped to preserve through restoration. Obviously, that one page turned into many, and brought with it an adventure that lasted more than a decade, and introduced many new opportunities. Discovery is a wonderful thing, through hard work and many long hours. Learning the names of those who had once called my house, their home, was exciting and satisfying. But more importantly, learning first hand about a special family, with a noteworthy story, from the lips of the last eyewitness to that history, was both a blessing and an inspiration. It was a turning point, as newly discovered doors of opportunity opened wide and dared me to enter. Truth be told, many of those doors had to be kicked in, as was often the case.

Too often the study of history has gotten a black eye, because it was written more like legalese, instead of the story of flesh and blood people. Too often, the recorded rigors of the past were long and dry, drawn out tomes of facts and figures that could be, in some cases, considered a cure for insomnia, rather than a personable story of how people lived and worked, and made their way through life.

While it is hoped that the information contained in this project will be both entertaining and informative, this final chapter can be likened to the "out takes" that are often tagged on to the end of a rented movie. This project was more than a gathering of information; it was an exploration of people's lives. Perhaps the way for the previous chapters to be best appreciated, is to share the process in which the information came to light, and the way in which this adventure unfolded, warts, wrinkles, and all.

It was February of 1990 when I took possession of the old house I used to walk past as a kid. The big old brick house was dirty and run down, the woodwork that trimmed the roof was rotting away. And despite her hard, weather beaten years, there was a beauty about her that shone through her blighted years. It wasn't until her cheap four-by-eight foot wall paneling and her drop ceilings were torn out, that her original beauty began to shine, for others to see. It would be many years, and many more dollars, before she was ready for a photo shoot from Better Homes and Gardens; don't call me, I'll call you when the paint is dry.

Filling the dumpster with old plaster turned into more of a full-time job, than my full-time job. And somewhere in all the dust and dirt, came an unexpected discovery, that would turn out to be the starting point of a most interesting journey into the past and beyond. Behind the last section of a plaster wall being removed were several snap shots, obviously they had been there for many years. The corners of this small stash of photographs were poking out of the plaster dust that had collected between the wall studs. They looked like tulips breaking through the ground in the spring. The photographs were about the same size as those I had seen pasted to the pages, in my grandmother's scrap book; it appeared that they had been taken with an old Kodak

Brownie camera. The find, at least to this writer, was more valuable than a pirate's chest full of gold coins, well, almost. With a fresh cup of hot tea in hand, it was time for a break, and a chance to study these new found images of the past I had just pulled from this time capsule of sorts.

In the first photo appeared four children, three boys and a girl; they wore hair styles of the late 1800's or early 1900's. Another photo was that of the same young girl, about eight, with a bow in her hair. Of course the first thought that came to mind was, "who are these people, and what do they have to do with this house?" Finding the photographs was the easy part, now all I had to do was find out who these people were, and why their pictures were stuffed in my wall. I spent the rest of the day looking for more hidden treasure in the walls of that big old barn, all that was found was more plaster and coal dust. Be it fate, luck, or destiny, those time lost photographs turned out to be first class tickets to a window seat on a time machine; little did I know I was already on board, logging frequent flyer miles. Time and opportunity would later put names to the faces in those photographs. If anyone on the planet had any answers about this old house, or those photographs, it would be Sis Johnston; she and her father were all that was left of the last single family to call the house home, before it was divided up into five apartments, during the Great Depression of the 1930's. All the writer had to do now, to get those answers, or at least ask the questions, was to get past the front door of Captain Johnston's daughter's home. Getting Sis to take the chain off the front door and allow me in took a couple of years.

Sis Johnston and I had talked on the telephone, many times, for a couple of years. Sis Johnston had also mailed several handwritten notes to the television station where I worked, with some not so subtle grammatical corrections to some of my "On Air" news stories.

I had visited her home numerous times, to hand deliver various items. I had to maneuver everything through the two inch gap, allowed by the door chain. I felt honored that I was allowed on the porch. On a humorous note, (and you'll certainly need a sense of humor to restore an old house) I had to wonder, if I were on the other side of that door, would I have let someone like myself in?

It was a sunny Saturday afternoon in early fall when I was finally granted permission to enter, and what a welcome sound that door chain made when it was unlatched. Sis Johnston was gracious but reserved as I came into her home for the first time; I had no idea what to expect, but I somehow knew it would be an experience I would never forget.

"Have you found my china doll young man?" that was how Sis Johnston greeted me on our first meeting. There was no doubt that I was on her turf, and that she was in quiet, but full control of the situation. Sis Johnston had spent her entire life in front of a classroom or coaching young athletes in their extra, after school curricular activities. Shaking hands with Sis was more like shaking hands with the varsity football coach than a gray-haired, retired school teacher.

According to some of her former students, "Miss Johnston didn't put up with any nonsense in her classroom. But, on rare occasion, she would crack a joke, and break up the entire class, all while keeping a straight face. When Sis Johnston smiled in class, everyone smiled," they said, "Sis Johnston was very old-fashioned, but very up-to-date at the same time." Sis was reared during Victorian times, grew up in a strict household, and was the only daughter of one of the most respected sea captains in the business. Captain Alexander Johnston's insistence on following the rules, rubbed off, or perhaps better said, were ingrained in everyone of his children. And as Sis has said more than once, "The captain gave us his orders and we would carry them out immediately, or else. In modern times the term is called tough love," she said.

Sis employed the same philosophy during her career as a high school teacher, as taught to her by her parents. "I loved my students, just like my mother and father loved me," Sis said. "There wasn't all the hugging and kissing going on back then that you see today, but we (children) all knew we were loved, and we all turned out O.K." Sis lived those times, and was a well-reared product of the times. The writer hopes to share not only the words of Sis Johnston, but also the energy Sis shared, during those many "visits" of ours.

Sis Johnston may have been enjoying her mid-80's, but she was anything but an old lady. Sis still drove her mid-sized Chevy to wherever she wanted to go. And it would appear that no matter how old we live to be, there are certain things that can still get us all riled up. As the story goes, Sis was headed back home from a monthly lunch, shared with some of her retired teacher friends, when her car was T-boned by a distracted driver who ran a red light. Sis's car was knocked off the road and onto the lawn of the Bay County Courthouse; Sis was ejected from the vehicle. When an officer from the Bay City Police Department arrived on scene, he was greeted by a woman, who was madder than a wet hen. Sis's car took a pretty good pounding and from the looks of things someone might be seriously injured.

In all of the confusion of getting hit by a car, tossed onto the courthouse lawn, and then thrown from her vehicle, Sis Johnston seemed to have a more serious issue to deal with at the moment. Sis had lost her wig, and wasn't at all happy about it. Anna Mae Wires, Sis's long time friend says when the police asked if she was alright, she told the officer to go find her wig and they would discuss the questions he had about the accident later. It seems the officer had once been a student of Sis Johnston at the High School and knew better than to argue with her, badge or no badge. It took a couple of moments, but the officer did find the wig, handed it to Sis and she immediately put it back onto her head and used the side view mirror of her crumpled car to return it to its original perch.

Here's a woman, lucky to be alive, concerned about how good she's going to look, after having her bones rattled in an automobile accident. Sis suffered bumps, bruises and a few scrapes; her car however didn't fare as well. But, the worst was still yet to come for Sis, at least that's how it seemed. When the accident was reported in the next day's newspaper, Sis, it was said, was less than happy. With her index finger, jabbing at the news print, right there in black and white, was not only Sis Johnston's name, and the particulars of her car accident, but sticking out like a sore thumb, for the whole world to see, was her closest guarded secret, now exposed, her correct age. Sis wasn't bothered with the fact that everyone now knew the exact number of years it had been since her home birth on Ninth Street, she just didn't think it was anyone's business. "Sis never took a shine to those imposing persons who had nose trouble." Sis did get over the incident, but it was those who were wise, who did not bring up the matter for discussion.

Sis purchased her first home computer when she was in her early 80's, so she could see what all the fuss was about. She didn't want to be left out of anything, and with a computer, she could e-mail her friends that were scattered across the country and around the world. Sis stayed on top of current events with the Internet, e-mail and her cable television.

Sis was afraid of nothing; she offered a chair and pointed to where she wanted me to sit. I was finally eye-to-eye with the sea captain's daughter, and the only person still alive who could tell me everything I wanted to know about the people in her family, about her former home, and about everyday life, back in those "olden days". I had many questions for Sis, and wasn't exactly sure which one to ask first. Alexandra Johnston had not only been born in the house I now call home, it's where her life and the lives of her family unfolded, and where her parents had grown old. It is the place where her mother and her brother Neal, both took their last breaths.

Sandwiched in between those events were the memories of several lifetimes and certainly many wonderful stories. Blessed to share the same room with Sis, I watched as the color in her face changed from a white pallor to a healthy pink, the depth of her aging eyes became bottomless pools as she prepared to share those precious memories, from the years gone by.

I sat on the edge of the old wooden chair and leaned forward, to hear her speak. But before she uttered a single word, she took a brief moment to study my face. After a moment of silence, the retired school teacher then gave me a look as if I had just told her, "the dog ate my homework," she leaned forward and said, "Have you found my china doll yet?" I wasn't sure how to answer the question. "What china doll," I asked myself, why was this woman asking me this question. Was this woman senile? Did she think, for the moment, that I was her father or one of her brothers? I was somewhat baffled by her question, and couldn't help but wonder if she had full faculty. "Could you explain the question for me please?," I asked. Not realizing it at the time, I had just been thrown back eighty plus years in time with Sis, in the wink of an eye.

Sis explained her question to me; she had lost her favorite china doll in the house, and never found it. She said, that her older brother Chuck had taken the doll and hid it from her, in the attic many years ago, as the two were playing hide and seek. And, as the gentleman who was now fixing up her old house, wondered if I had come across her doll yet?

Sis Johnston had taken me off guard, I was the one who came to the house to ask questions; instead, I was on the receiving end of a pop quiz, administered by the former school teacher. I wasn't sure how this adventure was going to play out, but I was quick to learn, the quest would be spearheaded by a china doll, on the lam, for more than

8decades. I had to be sure to squeeze in a few questions of my own, before I got sidetracked and went off hunting for a china doll that was AWOL. The doll had been a gift from her mother, and had been lost, about the same time the famed *Titanic* struck an iceberg and sank.

The moments spent with Sis Johnston were a treasure; she was the last person alive who had the answers to my long string of questions. Sis alone had the knowledge I needed, answers that could not be found in a dusty old book or in some forgotten file. It was Sis's secretary Anna Mae Wires, her long-time friend, confidant, and record keeper that finally persuaded her to let me past the front door, to see what I wanted. After all, "shouldn't we find out more about this fellow who has all the nose trouble?", Anna Mae asked.

Alexandra Johnston was puzzled as to why anyone would be asking so many questions about her father and mother or even her brothers? They had all been long gone for years, and what would this stranger want with any of them. "Why did this man want to get into my house, and why is it so important that so many personal questions be asked or answered? Who on earth would be asking such questions and what is he going to do with that information?"

Sis did not understand why anyone would be so interested in a bunch of people they didn't even know, and they didn't do anything special, that she knew of, to be asking so many questions. She looked at me and said, "Why do you ask all of these questions, are you writing a book?"

"Well, as a matter of fact Ms. Johnston, I am doing some research on the people who used to live in my house. I was told that you once lived there and would likely be the person to answer any of the questions that I might have."

"Who told you that," Sis quipped. "Who was it that said I had the answers; and that I would answer any questions about my personal life or my family?"

"Well, Ms. Johnston, it was a woman who said she worked with you over at the high school for many years. And she said that you were a very nice and wonderful person. "Sis didn't fall off the turnip truck last night; she raised her eyebrow and tilted her head slightly forward as if to prompt me to finish my "sales pitch."

"It was election clerk, Joann Vanover, who had taken notice of my new home address. After moving into your old house, I was assigned a new voting precinct, the same precinct where Mrs. Vanover works on every election. She seemed like a nice lady, and it seemed rather odd to me at first that she should care so much about my house and having me get in touch with you. Mrs. Vanover told me that the two of you often get together with some of the other retired school teachers for lunch, and that I could learn a lot from you. That's when she gave me your name and telephone number and instructed me to get in touch with you if I really wanted to know anything about the house."

It had taken a while, but I think Ms. Johnston was finally convinced I wasn't trying to sell her a vacuum cleaner, or a life insurance policy, and I didn't want anything more from her than to share some of her memories with me. After all, we did have something in common, we have both lived in the place I called home, the same place she too, once called home. I knew when she was able to relax, I could learn volumes from her, not only about the house and her family, but about many of the things she had witnessed in her years on the planet.

"I would like to write a short history of your old house and maybe find an early photograph of the place," I told her. "I want to frame them and hang them in the foyer and keep a little piece of history alive." Sis nodded and asked, "What would you like to know?"

That's when I pulled out the photographs that were found in the wall, and asked her if there was anything that was familiar in the pictures. She took the snapshots from my hand and studied each photograph, she nodded and said yes, she was familiar with those tiny pictures. "We had looked for these some years ago, where did you find them?" she asked. "They were discovered in a wall that was being torn out, I told her." She nodded, and then pointed to each face as she mentioned their names and saved the last identification, for herself, "that's me," she said. Before digging any further into her family's past, I had to know more about the captain's daughter before pressing on with this journey in time. Sis had moved about eight blocks down the same street where she had grown up; it was close to the school where she worked and the pump, (a well water hand pump, located on every other corner of her street) was close by.

She had lived alone in recent years after the death of her best friend and housemate Gertrude Butterfield. Gert and Sis had been friends since they were school chums. The lives of both women often paralleled and mirrored each others; they attended school together, they studied together, worked at Gert's father's photography studio, became teachers together and both cared for their aging parents until the end. Sis was still very much alive, still had many friends, and read her Bible regularly.

I had noticed something very different about Sis Johnston, while some folks of later years may begin to disregard world events, Sis was still quite interested in global affairs and kept up on the news of the day through television and talk radio. Although in her golden years at the time, Alexandra Stowe Johnston was still as sharp as a razor, and if she had something to say about anything, one way or the other, she said it. That, as I came to learn, was the beauty of dealing with the captain's daughter, you knew exactly where you stood with her, whether is was her good side, or the other.

Veteran Great Lakes Captain Alexander Johnston ran a very tight ship, at sea and at home. The good captain served as an example for hard work and discipline for many years to his crew and his children, the lessons he taught couldn't help but rub off on those under his command, especially his children. The slackers on Captain Johnston's crew could be fired, his children were quite another story. Sis remembered her father's lesson well, you have only to ask the thousands of former high school students who remember, her focused, no nonsense attitude and style when it came to the classroom. That she says, was a very important lesson she had learned from her father.

On my first visit to Sis's home I couldn't help but notice the prominent and elegantly framed, life-size photograph of a rather handsome gentleman sporting a large mustache wearing a gold watch chain strung across his wide chest. "That is my father! Captain Alexander Johnston," proudly proclaimed his only daughter. The gold charm that hung from the captain's watch chain caught my eye; I was drawn in closer to the huge stoic image for a more detailed inspection. Sis said her father had the photograph taken when he became a captain, in the late 1800's. "He used to work the Great Lakes and we would only see him during the winter." A captain? A Great Lakes captain? A sea captain used to live in my house? "That was my father and over here is a photograph of my mother when she was a young girl." Sis's mother had struck a classic pose, more than 100 years ago, wearing a long sleeved, black fitted dress with a large bustle that looked as if she stepped off the cobble stone streets of Victorian London. Her flower topped hat made her look taller and older than her seventeen years. The delicate black kidskin gloves Sis's mother was wearing reflected the life style of a well-mannered and rather well-to-do woman. "That picture was made from a tin-type taken in the late 1870's, my brother had several copies made from it," said the captain's daughter. It was obvious that Sis had conducted this home tour many times before. I could only smile as she spoke, understanding how lucky I was to be here for a personal tour.

What I originally had in mind was telling a short story of the house on a single sheet of paper for visiting guests to enjoy as they waited, in the foyer, for a door to be answered, or a taxi to arrive. It didn't take long before the writer realized this adventure was going to wind up being more than just a one minute time killer stuck on the wall in the foyer. I had become the benefactor of an historic informational windfall, and there was more, much more to come. Do I now simply carry out the task of writing a few paragraphs to frame and hang in the foyer, or do I jump in with both big feet, and see where this adventure takes me? And then what happens if there is enough good material to put a book together? I've never written a book before, and if I did, who would want to read it, and would I ever have the time to take the project seriously enough to actually sit down and write it? Those questions have answers, and at that moment, I wasn't sure where I would find them.

Promising to look into the issue of her missing china doll, I began a respectful interrogation into Sis Johnston's past. This, I was quick to learn, was not a television reporter questioning a politician. This was a student dealing with a savvy teacher, a teacher thirty years removed from the classroom; she was sharp and was still in charge.

I had asked Sis another question about her father; there was a very long pause, Sis didn't say anything, and I wasn't sure she had heard the question, and so I asked her again. "I heard you the first time young man! I was thinking about the last question you had asked me, now if you'll give me a moment, I'll answer your first question!" "Yes ma'am, I said." "Don't call me ma'am, my friends all call me Sis, you can call me Sis too."

Whew, I thought I had blown it for a minute there, it took me three years to get in the door and only 10 minutes to nearly get myself quickly ushered out the same door for insubordination. It had been a long time since anyone had asked Sis specific questions about her family and I knew by the look in her eyes that there were many emotions attached to the memories that were rocketing to the surface after so many decades of sealed, quiet slumber. After all, who was still alive that personally knew her father or mother or any of the neighborhood kids that knew her brothers. "There were brothers Frank, Charles, Neal and myself," Sis explained. "Frank and Charles did well for themselves, in business, on the East coast." Frank made his fortune in New York in the advertising business after he sold his interest in his grocery store in Michigan."

Charles worked for the U.S. government in Europe during the second war. Charles, as the family tells the story, was a civilian one-day, and a soldier in uniform, the next. Charles had been commissioned as a colonel in the Army, working out of the Office of War Information. Sis wasn't sure what her brother's job in the service was, but he held such a position that he was allowed to censor his own overseas letters, he would send back home from Europe. Charles Johnston's children say they can recall living overseas with their father and attending black tie parties in England that were often held in castles and large houses where many heads of state and other VIP's would attend.

"Frank Johnston was a successful business man and bought a mansion outside of Washington D.C. that was built in 1803. His property was of historical significance, and really something to see," Sis said. The estate was called San Domingo, by all appearances it was an old time plantation. Living at San Domingo, according to her oldest brother, was the way life should be lived. San Domingo often appeared in photos on the society page in the local Washington D.C. newspapers. Frank's wife had a wonderful garden and a wonderful home, and they raised Black Angus beef in the back forty. Actually, there were hired workers to take care of the ranch. The meat, according to Sis, "was better than wonderful." Sis's father was a ship's captain, her brothers, one a millionaire advertising executive and the other a colonel with a secret military job during the second world war, were sure to be great stories within themselves. However, my focus remained on the retired school teacher, and her story.

My first face to face meeting with Sis Johnston had roared past much too quickly; it seemed like only a few moments, but several hours had slipped by and I still had a long list of questions left to ask her. I could tell Sis had quite a work out with my non-stop inquisition; she was getting tired. I thanked her for her time and went home to re-explore the house from top to bottom, armed with a handful of clues as to what I should be looking for, and absolutely no idea as to what I might find.

It had been more than sixty years since Captain Johnston and his daughter Sis had lived in their stately home on Ninth Street, and nearly eighty years since that china doll was lost during a game of hide and seek with her brother Chuck, in the forbidden attic. I was left to wonder, if somewhere in that house, there wasn't still some sort of evidence left behind by the Johnston's, perhaps something hidden or lost. The only untouched areas of the house had to be behind the plaster walls and maybe under the floorboards in the attic. I had convinced myself that if there was something to be found, I was going to find it, all I had to do was keep my eyes peeled.

The old brick house had changed hands at least a half dozen times, at the very least, after Captain Johnston and his daughter sold the house and found a smaller, more manageable apartment nearby. The two story brick, Italianate Victorian home had an elegant but subtle demeanor; it had been converted over to apartments during the depression of the 1930's and was home to many dozens of apartment dwellers over the years, so was it still possible there still might be something left behind by the Johnstons so many years ago?

An easy find was in the attic. A couple of the house's original shutters were stowed away, still intact, and apparently had been stashed away sometime during the 1960's when inexpensive aluminum storm windows had been installed. There were plenty of other remnants of the past too, also found in that attic. There were shovels full of evidence that pigeons once had reign of that section of the house; a putty knife, a stiff wire brush and my Sears wet/dry industrial vacuum took care of that. The attic floorboards were an interesting study in themselves. The rough sawn pine boards were a full one inch thick and measured between 18 to 24 inches wide, and anywhere from 15 feet or longer. The dimensions of the lumber may be of no consequence to many, but to wood workers and restoration people, the floor boards were not only a welcome find, but also conjured images of what could be created with these well seasoned antique materials. A number of those floor boards were nailed down with square nails. And quite a number of them had apparently been pried loose over the years to run telephone and electrical wires, and this is where the house gave up her first secrets, a small booty of treasure. As simple as it may be and as strange as it may sound to some, turning over an old floorboard up there was almost like letting the Genie out of her bottle. This house was now ready to share some of her time locked secrets with me.

The first discovery was a horseshoe wrapped in yellow velvet, a child's building block, and a most peculiar contraption made from a wooden stick with a wooden star wheel, on the end of it, that rotated. Could it be possible that these things once belonged to Sis or her brothers? Was it possible that these things were hidden away and forgotten, or were they lost, waiting to someday be re-discovered so they could help tell a story? Or were these things just junk that may have fallen out of storage boxes, stowed away by countless renters who have come and gone over the years?

Only a few days had passed since I had met with Sis. I did not want to become a pest, but I had to go back, to find out if these attic treasures, covered with years of coal dust, were in any way familiar to her, or just some junk left behind by who knows who. Confirmation would require another face to face meeting with the retired school teacher. Would she make me wait another three years before she would let me back in the front door? Had she had enough of me, and my questions, from the last visit? There was only one way to find out, so I called her on the phone. "Sure, c'mon over," Sis said, "I'll have some coffee ready and we can chat some more."

When I got to her place, Sis was resting comfortably in her easy chair, where she could keep an eye on the television, the front door, and the coffee pot all at the same time. It was hard not to both enjoy and be amused by the look on Sis's face when I walked through the door with my wrinkled old grocery bag full of goodies. There was a certain twinkle in her eye, and although she attempted to remain reserved, composed, and slightly disinterested, I could tell she was very excited about the possibility of rediscovering a few pieces of her very distant past. If in fact any of these items found in the attic were hers or her brothers, it would have been seventy or eighty years since she had last seen them. Would she recognize any of this stuff in the old brown bag, could any of it have belonged to her or her brothers, and if she was familiar with any of it, would she really care? Anna Mae Wires, Sis's assistant, poured the three of us a cup of coffee. I had hoped to sit quietly by and let Sis do all of the talking in hopes I could learn more about her, her family and her father, the captain. Instead she began the conversation with several questions, "Have you found my doll yet? What do you know about the Myron, and how long are you going to make me wait until you show me what's in the bag?" While it had been nearly 3 decades since her retirement, working as a teacher, Sis Johnston still had that certain something that could make you sit up straight and pay attention. I felt as if I had a homework assignment that was due immediately and the teacher was impatiently waiting for it. Actually, it was a homework assignment, and I was ready to see what kind of grades I would be getting from the teacher.

The first item I pulled from the bag was a steel horseshoe, wrapped in gold velvet ribbon. I had never seen anyone wrap a horseshoe in any kind of ribbon before and thought that this item had to be special in some way. Before I had a chance to ask Sis about the shoe, she pulled it from my hand and smiled. She said, "This is from one of my school plays, and it's still around!" The child's building block was no stranger to her either, "That was mine!" Then it was time to pull out the stick with the spinning star wheel. Sis gently took the hand made child's toy without saying a word. She slowly surveyed the curious wonder, from both sides, and from end to end. A few quiet moments later Sis looked up at me, over the lenses of her glasses, she had a one-word response, "Chuck!"

"This was Chuck's, he had a wonderful sense of humor, and a bit of the devil in him from time to time, and look here it still works, it still makes noise." Sis said, "it was the kind of toy the boys would use on the neighbor's windows that would make a lot of racket and would shake them up inside." It was also the kind of toy that was especially popular during the Halloween season when the lines between mischief and fun became blurred for a boy with too much time on his hands. Sis said, "Chuck would take a dare from his friends and run up on the neighbors porch and clunk the little star wheel across the living room window, it would make a terrible racket inside the house." Then, Chuck would run away and hide in the bushes and wait for someone inside to scramble out onto the porch to see which neighborhood hooligan was causing all of the trouble. "I think it was Uncle Walter who taught the boys that trick, and a few other tricks as well," she said.

I did get the star wheel stick back from Sis but I never did see that golden horseshoe or the child's block again, nor was I interested in asking for them. Sis seemed very pleased and quite happy with the discovery of a few of her childhood mementos; I would try to put another smile on her face by hunting for more of that lost treasure.

With each visit, Sis Johnston became more relaxed, and began to share many more of those little bits and pieces of what life was like, when she was just a young girl growing up on Ninth Street. After a few quick hours, Sis said, that was it for today, and that I was welcome to come back to her home for another chat, but right now she was busy, very busy. I found myself being quickly ushered out of the house. Before I said goodbye to Anna Mae at the front door, I turned to see Sis pick up her remote control for the television. The TV commentator said, "It's the University of Michigan Wolverines on the air!" It was only moments before the kickoff of a Saturday afternoon U of M football game, and one of the Maze and Blues biggest fans was settling in for the gridiron showdown. Sis loved her Wolverines.

Anna Mae said, "when Michigan is playing ball, Sis is totally focused, she won't take a phone call or even take her medicine when they're playing." I thought to myself, "Go Blue." The University of Michigan is where Sis earned her Masters Degree, and even into her late eighties, a dyed in the wool Wolverines football fan.

As I left the porch of Sis's house it was clear, that if I was going to do things right, I would be writing more than a few paragraphs on the history of the Johnston's old house, but where the trail would lead, and how much the adventure was going to cost me, in terms of time and dollars, was still unclear. It appeared that I had unlocked the proverbial Pandora's Box. Now, I would have to start looking for that china doll, and any other evidence that could help tell the story of an old house and the people who once lived there.

There were some decisions that had to be made; as the property owner, should the house be turned back into a single-family residence or kept as the apartment house it had become after the captain and his daughter moved out in the early 1930's? The decision was based more on need than want, to keep the lights on and fund the restoration, you have to have cash, and lots of it. An apartment house it shall remain, at least for the foreseeable future, while the project remains underway.

In the coming years the house would swallow up many dollars, and then some. There were long years of hard work ahead, and the project became more of a treasure hunt of sorts with each wall that was torn out for new wiring, plumbing and insulation. While little more than coal dust and plaster chunks were found behind the walls, there was still the chance that "something" else might be discovered, even if it were only a scrap of an old newspaper or a discarded beer bottle.

Much of the yet to be discovered treasure would later be found in piles of old papers from Sis's personal files, that were earmarked for the garbage heap.

Sis Johnston had defied time and the odds into her later years; she still climbed the stairs, every night, to her second floor bedroom. But an increasingly troublesome knee and age were starting to catch up with her. The time had now come for smaller more manageable quarters with no stairs to climb. In the process of moving, Sis's collection of several lifetimes of treasures and memorabilia had to be dealt with.

Fortunately, this writer was lucky enough to be at the curb, only moments before the junk man was scheduled to show up with his truck. I was at the right place at the right time, and here was Sis's trash, all mine for the taking. Sis had given her personal assistant, Anna Mae, strict orders, "To clear out that old stuff in the file cabinets and find a home for those other things." I found myself under the watchful eye of Sis's protective neighbors as I rummaged through each "Hefty" bag, that had been tossed to the curb. 99-point-9 percent of the material, was in fact, trash, but there were a few items discovered, that to me, were treasure. I was looking for old photographs and anything else that would help tell the story of the Johnston family. What I came away with were several yellowed newspaper clippings and a few more pictures, that were taken of her brothers. "That's the stuff Sis wants me to toss out, she doesn't have room for it anymore and who else would want it?" asked Anna Mae. Sis's hand held magnifying glass was where she had left it; on her side table. I used it to study the tiny old photographs. As the yellowed snapshot came into focus under the magnifying glass a most interesting image appeared. I could now distinguish the features on the faces of those three boys and that little girl, lined up from tallest to shortest. Could that little girl on the end be Sis? And could those other guys be her brothers? And could the photo have been taken by her father, the captain, aboard one of his ships? The answer to all of those questions was yes; Sis confirmed that her father had in fact taken the photo, in 1915, using a box camera. Nettie Johnston, her sister, Aunt Mae Neal, and the children came to visit father for a few weeks on board his freighter the *Powell Stackhouse*.

There were a handful of other photos, that became stuffing for those huge "Hefty" bags that piled up at the curb. Was it luck, or was it fate, that put this writer, in the right place at the right time to become a junk picker? And to think, in one afternoon, finding so much treasure moments before the trash man and his truck made it to Sis's house.

Now realizing my dilemma, I had far too much information for a simple one page history of the house to be hung in the foyer; I now had the beginnings of what appeared to be taking on the shape of a short story, perhaps even a book. By now I was hooked, I had to know more, I had to see what else I could find, and I believed I knew where I could find it, or at least where I should start looking.

With more than a few years under my belt working as a television news reporter, gathering information was what I did best. But I was left to wonder just how tough could it be to dig deep, into the past, to come up with more information on Sis's old house and learn more about the people who once called it home? The tough part of this job, was to listen to everything, every minute detail that came from Sis's mouth. She may have been reserved at the start, but now, she was remembering things, she said, she hasn't thought about for years and years.

I had been paying particular attention to Sis's father, mother and her brothers, thinking, at first, this was the story unraveling before me. But as I listened to the stories Sis had to share, it became more apparent that this adventure was focusing more often than not on Uncle Walter's influences and his adventures.

Dubbed the "Iron Man" for surviving 20 hours in sub zero temperatures on Lake Superior, following the wreck of the *Myron*, Captain Walter Neal is very much a part of the lore and legend. Fifty-six years after the loss of the Myron, and not far from where she sank, another ship's crew would find themselves in peril on Superior, and like the crew of the *Myron*, would lose their lives to the Witch of November. In 1975, the Great Lakes freighter, *Edmund Fitzgerald* would go down with all hands, but unlike the *Myron* tragedy, not one body from the *Fitzgerald's* crew was ever recovered.

Uncle Walter, as it turns out, really was the stuff legends are made of. So, as I listened to story after story from Sis about Uncle Walter, it wasn't at all difficult to understand how that special man, earned a special place, so near and dear to her heart, and kept it, even so many years after his passing.

Uncle Walter was like many folks who enjoy a good homemade stew, a good joke, a long smoke and if the need be, to bend the rules, just a little bit to get by. Uncle Walter not only bent the rules, he broke more than a few of them, more than once, along the way. Sis's special uncle was just the opposite of her father. While Sis was not entirely specific

on the details of the rivalry between her father and her
uncle, it does become quite clear over time that the two
captains, each had a different take on how life, business and
family were to be run. Captain Alexander Johnston was ten
years older than his brother-in-law, Walter. Alex had served
as first mate under Walter's father, Captain William Neal
and was a heavily starched and dyed in the wool Republican.
Captain Johnston, according to his only daughter, lacked
for a sense of humor, walked with a swagger of arrogance,
and was a product of his time and environment, truly a man
of the Victorian age, with all of the rules and protocol that
were fitting and proper for the time.

In tribute to her father, Sis had hung on to a prized
wooden chair she said her father had owned from the time
he lived at the old Campbell House, before he married Nettie
Neal. The Campbell House had been what we know today
as a hotel/conference center of sorts, with barbershops and
restaurants and other outlets of professional necessity. The
Campbell House hotel was known in the 1880's as one of
the nicer places to live, but was also located on Water street,
known at the time as "Hells' Half Mile." A short, but rowdy
stretch of Water Street that was home to numerous saloon
fronted brothels that served as a magnet for lonely hearts,
sailors and lumberjacks (shanty boys) alike. A robust business
district, and center point for travelers in the daylight hours,
Water Street wore a different hat under the night sky. Sinful
and disgustingly rowdy was just one way to describe the area.
Wild is what it became in the winter with the sailors in port,
and wild is what it stayed when the sailors shipped out and
the lumberjacks were back in town after the long cold and
lonely winter in the woods, falling trees.

It is, however, still unclear whether young Captain
Johnston procured his chair from the Campbell House while
a resident, or whether the chair was a family heirloom he
brought with him from Canada.

"The Captain's Chair" / photo David Hetzel

The captain's sitting chair was a familiar fixture in the Johnston home; it was often offered to guests when they came to call on the captain. In the early 1930's, after the passing of Sis's youngest brother Neal, the chair would again follow the captain to his new quarters after selling off his Ninth Street house. After Captain Johnston's passing in January of 1941, the captain's chair became an heirloom that would remain under the protection of his daughter, who over the years, recovered the chair several times. The decades rolled by and almost three quarters of a century later, the chair has returned to the captain's home, where it again sits in its original location. At times the chair was used near the fireplace where the captain himself would warm his cold hands. In the warmer weather, Father's chair sat next to his big roll top desk in the library, he often used the chair near the window to do his reading. "The writer was honored that Sis would bestow the stewardship of her father's chair; after all, this is the same woman who would not let the man with all of the "nose trouble" past her front door, for several years, before finally getting permission to enter. Even after so many years, the captain's "Eastlake" style chair was still solid, and is comfortable, or at least, Victorian comfortable. The decorative carving was simple, the arms were graced with short, turned spindles and it was certainly wide enough in the saddle to declare it was a man's chair. The captain's wooden relic, is back home.

As Sis settled into her new, more manageable quarters, many of the things she had collected over a lifetime were being passed on and put into the hands of a few close friends and several area museums. The life-size photograph of her father, in its elegant frame, is part of the permanent collection of the Saginaw River Marine Historical Society. Many other family and maritime artifacts, from her grandfather William Neal, her father Alex Johnston and her Uncle Walter Neal have found their way to area museums.

The few known artifacts that remain from the wreck of the *Myron* can be found in several maritime museums in northern Michigan, at Sault Ste. Marie, at Whitefish Point, and Rogers City. The wood and iron anchor, retrieved from the *Myron's* watery grave site, is on display in the hold of the permanently moored and retired freighter *Valley Camp* which lies just east of the Soo locks. The *Valley Camp*, built in 1917, claims to be the largest Great Lakes maritime museum, and features numerous exhibits from shipbuilding to shipwrecks. The keepers of the retired coal fired steel relic, allow visitors to come aboard to spend as much time as they wish to wander the vessel from stem to stern, above and below deck to get the feel of life on the lakes. *The Valley Camp* is also home to the torn and twisted, wave battered lifeboats of the *Edmund Fitzgerald*. The power of the lakes fury on the day that took her crew's life has been permanently recorded in each rip and tear of the yawls thick aluminum hulls. The "*Fitz's*" life boats never fail to steal the breath away from the thousands of people who re-explore the November 1975 tragedy. The recovered wreckage still bears the weathered name of her mother ship. The personal eyewitness inspection of the carnage, brings with it, an aching heart, and a long moment of silence. Visitors pass in stoic reverence as they imagine the last terrifying moments aboard the *Edmund Fitzgerald*, before she was so violently swallowed by Superior.

A must stop for maritime family members, history lovers, and the curious, is the Great Lakes Shipwreck Historical Society's Museum, at Whitefish Point, Michigan. There you will find more than 150 years of shipwreck history, and restored an 1861 lighthouse facility from the lighthouse itself, commissioned by President Abraham Lincoln, to the numerous buildings that once housed keepers, crew and boats.

One of the museum's crown jewels, is the actual bell of the *Edmund Fitzgerald*. The Historical Society reports that, on July the 4th of 1995, at the request of the crew members surviving families, the ship's bell was brought to the surface. The bell is now on permanent display in the main museum building at Whitefish Point. At last count, nearly ninety thousand visitors, make the trek, and tour the lighthouse, museum and the deadly shoreline, of Whitefish Point, each year.

Just a whisper away from the brass bell of the *"Fitz"*, was the reason this writer crossed the five mile long, Mighty Mackinaw Bridge, and made the trek to Whitefish Point. The *Myron's* poolished brass wheel, stands in marked contrast to its deep blue backdrop. The wheel, stands in silent tribute to her crew, lost in a late November gale, on Superior. The tiny wooden steamer was beaten to the bottom, more than 50 years before the *"Fitz"* disappeared from the surface. The souls of 16 godforsaken sailors were lost when the *Myron* took her last breath and slammed to the bottom of Superior. Her skipper, Captain Walter R. Neal, was holding that very wheel for dear life, in water up to his knees, when a wave tore the pilothouse from the *Myron's* deck. Uncle Walter, was her only survivor, rescued after twenty hours in the drink, and lived to tell his story. According to his niece, Sis Johnston, Uncle Walter said a prayer every night for his men, until the day he died, more than thirty years later. Sis Johnston said, she has gotten her feet wet, more than once, standing on shore at Whitefish Point; saying a prayer for every man lost on Uncle Walter's boat; and for those on the *Fitzgerald* too. Sis said sailors are special people, and she was keeping close tabs on the new museum that was all about the men who worked on the ships. Sis was especially interested, not only because she thought Uncle Walter should have a place there, but because the man organizing the collection, is a retired educator, like herself.

The blossoming Great Lakes Lore Maritime Museum in Rogers City, Michigan, was instituted to honor the men and women who have braved the inland seas. There you will see the faces of the merchant mariners and learn about their lives aboard the ships of commerce. Their uniforms, their tools and the instruments used to navigate the lakes, are also on display. Museum Director Ed Brklacich (pronounced Burr-kla-vich) originally established the museum in the town of Sebewaing, Mi. on the southern shore of Saginaw Bay. Brklacich, is a retired secondary school educator, whose thirst for knowledge did not end with his retirement. At the close of each summer, the Great Lakes Lore Maritime Museum inducts a small group of mariners, into its historic family.

Many artifacts and mementoes from the yesterday's of merchant sailing, remain in the caring hands of the ancestors of sailors, who have passed. Those privately held treasures are now used, to tell the stories of the old sailors to the great grandsons and granddaughters so they may pass the stories on to their children and grandchildren. But many historians say, too many of those artifacts, are getting lost, or are being sold off by descendants, who are not familiar with the stories that accompany the once treasured relics.

That is not the case for the heirs of Captain Alexander Johnston. The trusted pocket watch of Captain Johnston, his sea chest, and a writing desk used aboard one of his early vessels, were handed down to the captain's great-grandson, Holt Johnston. Holt said, "he keeps his great-grandfather's pocket watch in working order, protected in a safety deposit box so his children, and their children will be able to enjoy an important piece of their family history in the years to come. The pocket watch, according to Johnston, comes out of the safe several times a year to admire, share stories and to keep tabs on the inner workings to make sure it keeps, keeping time. The captain's pine sea chest, with its dovetailed corners, and his desk, have become household furnishings in his great-grandson's home for both family and friends to enjoy. A handful of other shipping related items, which relate to the Johnston/Neal family legacy, have found their way into other Great Lakes area museum collections, or the hands of close family friends.

Sis Johnston served as a life long steward for many of her family's treasures including a number of her father's personal things. There was one item, in particular, that appears to have been extra special to the captain's daughter. It was the same Masonic charm Captain Johnston wore so proudly in the photograph he had taken, upon becoming a commissioned captain, and achieving the 32nd Degree, as a Mason. It was special, not only for its sentimental value, but also, in part, for what Captain Johnston told his daughter before his passing. "Hold on to this Masonic charm, keep it with you always, if you ever need help for anything, take this to them and they will help you." Sis said, "He told me that a long time ago."

Alexandra kept the captain's charm with her, where ever she traveled. When she wanted to be close to her father, she would hold it in the palm of her hand and say a prayer. Knowing she wasn't long for this world, and knowing she would not be taking anything with her when she passed, Sis surrendered the charm, and her jewelry, as she prepared to move to her new home in assisted living.

Anna Mae Wires, was Sis's best friend, and it was Anna Mae, who Sis commissioned to take care of all of the details of her move.

Towards the end of the task of clearing Sis's home, Anna Mae had found herself left with a menagerie of odd items, that included a box of earrings that had long since lost their mates, along with bits and pieces of broken neck chains, bracelets and lapel pins. Anna Mae had contacted a local coin, gem and precious metal dealer to see if she could sell off those odd items, at Sis's request, for scrap. Sis said, she could use the profits for pin money. What I was not aware of, at the time, was that Anna Mae had taken my suggestion, and used the local coin and precious metal dealer I had recommended. He was honest, and he would take the time to explain each and every question I had when it came to buying old coins. The guy was easy going, and had a funny name, which made it very easy to remember.

John DeJohn began his coin and jewelry business as a vendor at a flea market and did so well he rented a storefront and did so well there, he bought the building. Mr. DeJohn wasn't a hustler, just a nice guy, who was generous with his knowledge, and his valuable time.

I had tried to be kind, and gentle, and not too overbearing, but I had been like a monkey on poor Anna Mae's back as she dealt with Sis's personal affairs. I kept asking if she had seen this item or that item and especially the charm that had hung from Captain Johnston's watch chain in his photograph. Anna Mae was a senior citizen herself, with more than plenty to do, and I didn't want to be too big of a pain in the stern. But I was on a mission, I was searching for hard evidence to go along with this unfolding saga of the captain's daughter, her family, and my house. How fortunate I was, to not only get some great leads on researching the history of the house, but to actually have the captain's daughter, willing and available, to share with me, the many stories of the events she had witnessed, with her own eyes.

I was more than interested in the historic value of the keepsakes that Sis had kept safely tucked away, over the years. I wanted a chance to touch and to hold a piece of Captain Johnston's personal past, something he had once owned, an item that he obviously treasured. And besides, the search for such an elusive prize would certainly bring with it, not only the thrill of discovery, and some new information, but tangible evidence, of a man who years ago, proudly called my house, his home.

While I hadn't given up hope on finding such a treasure, the chances of finding the captain's Masonic charm were getting slimmer as the weeks and months went by. By this time, Anna Mae had Sis's things at the house taken care of, for the most part. One day while having coffee with Anna Mae and her husband Larry, at their kitchen table, we were thumbing through Sis's leather bound family photo album, studying the photographs, and reading the notes written on the back. Most of the images were taken back in the 1870's and 1880's, some contained written information on the reverse side, other photos were not labeled and the three of us had fun trying to guess who might be related to whom. But somewhere between that second or third cup of coffee, and enjoying each other's company, it was something Anna Mae had casually mentioned in passing that set off a loud ringing alarm inside my head. She very softly said that, "all of that left over, odd ball stuff, went to your coin dealer buddy about a month and a half ago."

It was then I realized that if the captain's charm were going to be located, in this lifetime, it would have to be in the hands of honest John up at the coin shop, or it was probably lost forever. But this was a Sunday afternoon and John wouldn't be opening his store until tomorrow, and I knew I couldn't wait that long, I'd have to find him. I finished my coffee at Anna Mae's house and thanked her and Larry for their time. We've all heard stories about people who have won the million dollar lottery, and found the missing ticket, just hours before it expired. I now know what that experience feels like. I was a man on a mission, and I wasn't going to quit until I had that elusive prize in my hands.

Now, with a full tank of gas and like a man possessed, I had to track down honest John and get him to open his store. I knew of several places he liked to eat, and a few of the places he liked to hang around; I spent the next few hours combing the area for his car, and found nothing. With one hand on the wheel, and the other digging through the glove box, I did manage to dig up an old business card he had given me, a few years ago, with his home and cell phone number on it, but no luck. Finally, I left a message on his home answering machine and hoped for the best, as I cooled my heels.

It felt like I had just put a message in a bottle and thrown it out to sea, thinking what were the chances of that bottle being found. In the meantime, I had parked my car in front of his store, thinking maybe he would stop by to collect something he may have forgotten. After some time, a police officer pulled up behind me, and came over to ask why I was parked in front of an empty store for so long. I never thought of myself as a suspicious character before, but I guess it could have looked like I was casing the joint. After a brief chat with one of the city's finest, I started the car and drove off.

The phone rang not long after I got back home, it was John, and he seemed rather puzzled that I would call him on a Sunday.

"You want me to what?, he said. Open the store, why? I'm just sitting down to dinner!"

After I had explained to him that the treasure I was searching for may be stashed away in his safe, he agreed to meet me at the shop once he was finished with his dinner. I headed for the coin shop, to wait for John. As I pulled up in front of the store, I spotted that friendly police officer in my side view mirror as he happened by again, keeping a close eye on my Chevy, and John's store. Turning my head from the street, to the coin shop, I could see that John was already in the safe, and the front door was unlocked, he had spread several dozen small boxes of oddball jewelry across his glass topped counter. The booty, was neatly tucked into little ziplock bags, they were all labeled with the names of the previous owners. After many months of searching and asking questions, I was like the proverbial kid in a candy store. I had taken a blank check with me, just in case my hunch that he might have the captain's charm was on target. John gave me a courteous hello as I entered his store, his nose and his fingers were buried deep into his box of treasure.

He said, "Oh here's the bag I'm looking for, now let me dig down and see if there is anything in here that even remotely resembles the charm in that photo you have." John rustled around in those little plastic bags for what seemed like a lifetime.

"Well, by gosh and by golly," John said, "I think this might be what you are looking for!"

"Holy smokes John, that's it, look!, It matches the charm in the photo, look if I put it next to the charm in the photo it matches up exactly. John let's talk turkey here, you know how much I want this charm; to me, it's almost like shaking hands with the captain himself, so name your price." I was so busy being excited, with the discovery, I didn't notice that John was digging deeper into that little bag. My focus was on Captain Johnston's Masonic charm, and here it was, finally in my hand. "John name your price, I have to have this for my research."

With a very big grin on his face John said, "well I guess you're not interested in the captain's Masonic ring that has his name inscribed on the inside of it, are you?"

"Holy smokes John, lightening has stuck twice in the same place, after all of this time looking for the captain's charm you have his ring too?" This was just not to be believed, the charm was in one hand, and the captain's inscribed ring was in the other. Well, I hadn't played this one well at all; John knew I wanted the charm, and the ring, so now he had me where he wanted me.

"O.K. John, let's get down to business; now I know I took you away from the dinner table and I know that it's Sunday night and you could be home with your feet up watching the TV. But, you know I have to have these things, so how much?"

John didn't say anything for a few moments, then with a slight grin said, "You know, one of my regular customers who is a collector, of these types of items, he has been here over the last couple of days looking at that Masonic charm, but he thinks I want too much for it."

I though to myself, "oh great, not playing my cards close to the vest will certainly cost a very pretty penny."

John said, "buddy I'm going to make you a deal." I said, "John you're not going to rob me, are you?"

He let out with a hardy laugh, and then chuckled saying, "here's the deal, I know I could take you to the cleaners on the value of the scrap gold weight alone on both the charm and the ring, but I'm not going to do that. We're going to have a partnership, if you agree to the terms."

While John certainly didn't seem like the gangster type to me, I braced myself for the offer I couldn't refuse. I leaned my ear closer to him, to hear the terms of this so-called partnership.

Mr. DeJohn said, if I wished to buy the items outright at market price he would allow me to do that. The other proposal he made, after roasting me on the spit for a while, was almost too good to be true. And for that reason alone I became quite skeptical. But I had already decided that I was not going to leave his store without the captain's charm or the ring. John proposed that both of us could be equal partners in the ownership of my newly discovered treasure. "Well John, I am listening, go ahead and tell me about this proposal, you wish to make."

John became very business like and began to speak, "For one dollar, you and I will both own these items. If you donate them to the museum, I get half of the tax write off value. If you die, I get the charm and the ring back and if I die you get full ownership, is it a deal?"

"That sounds great John, but how are we going to handle this custody issue, do I have the charm and the ring every other weekend or what?" I asked.

John laughed so hard, tears came to his eyes, I knew he was having fun with me, but he also knew I was waiting for the other shoe to drop.

John said, "You will have full possession of both items, besides I know you're not skipping town anytime soon. So, is it a deal or do you want to pay me full price for that stuff?"

We shook hands, I gave him a dollar, John wrote up the partnership agreement, and I left the store with treasures in hand.

This was a moment to be savored, these new found items represented a small but wonderful victory of sorts over father time, and the odds. I couldn't wait to see what or whom the next discovery would bring, I knew this was not the end of the journey; all I had to do was keep digging and most importantly, pay attention.

It didn't take long before the next discovery would present itself, although I didn't know it at the time. What I had come to understand in this endeavor, was that when opportunity knocks, it sometimes does so, very softly. This next turn of events only underscores this theory.

This hard-working television reporter, who spent his spare time restoring an old captain's house, also needed to get out and relax from time to time. And it just so happened, this reporter, had the occasion to do a feature profile story on a local rock and roll legend, who had a number one hit record in the 1960's.

His name is Bobby Balderrama, lead guitar player for Question Mark and the Mysterians, whose number one hit "96 Tears," knocked the Monkey's, number one song, "The Last Train to Clarksville" out of the national top spot on the American pop hit parade. Robert Lee Balderrama was only 15 years old at the time his record sold a million copies, and he has the gold record on the wall, to prove it. But that part of Bobby Balderrama's life is history now. He's matured and moved on, and finally created his own style of music after being in the background of the band's mysterious lead singer, Question Mark, for more than thirty years. Balderrama was now his own front man.

The veteran guitar wizard, known for his skill and blazing lead guitar licks, was in the process of completing a new CD, with his new band. One of the featured songs of Bobby's new album was a remake of "Bluest Blues," originally recorded by British guitar guru Alvin Lee, of "Ten Years After" fame. Balderrama, an exceedingly accomplished guitarist, raised the bar on the 1960's blues song. The Robert Lee Blues Bands' version of "Bluest Blues" is one of those rare songs you hear on the radio, that carries you off and makes you forget where you're driving to. Balderrama touches the soul, as the bending strings of his guitar wail with the pain and heartbreak of coming home to an empty house, after losing a life long lover, to another man.

Later that night, after interviewing Bobby about his new CD, and his band, I was busy with chores and decided to listen to my promotional copy of "Bluest Blues" to make the job of doing the dishes, go by a little quicker. Somewhere after the third or fourth time "Bluest Blues" had played on the stereo, I found myself dusting off the guitar that sits in the corner. A couple of hours had gone by, the dishes still weren't done and all I wanted to do was play along with the band in the living room. "Bluest Blues" was now on that list of all time favorites. After Bobby's television interview was broadcast, Chad Cunningham, president of Bullfrog records, called to say he liked the story the photojournalist and I had put together. He added, that Bobby did a nice job with the interview, especially for being a shy person. Mr. Cunningham asked how I liked the song "Bluest Blues," and without thinking, I told him it was missing a guitar track. I explained that I had jammed with the song, on CD, for the last few days in my living room; he seemed somewhat amused by all of that, and chuckled over the phone.

To make a very long story short(er), it was a twist of fate, and a dare from Cunningham, that put me, and my Fender Stratocaster in the studio with the Robert Lee Blues Band, for an audition of sorts. I had goofed around with the guitar since I was a kid, but had never played in a real band before. I was shaking in my boots as I stood in the recording studio surrounded by four veteran musicians who had at least 30 years experience each, under their belts. Keyboard man, Frankie Rodriguez, was another original member of Question Mark and the Mysterians; Frankie was just 14 years old when he recorded "96 Tears." Bassman Ron Dozier, was from New York; he had recorded and jammed with some of the best Rhythm and Blues recording artists in the business. Ronnie was not only an accomplished bass guitar player, he was like greased lightening, with soul, on the six string guitar as well. Drummer, percussionist, songwriter and funnyman, Randy Hurry had been playing the drums since he was a kid. Randy knew and played with some of the best rock and roll notables on the West Coast for years. And here I was, ready to show these seasoned veterans, everything I didn't know about playing rhythm guitar. But I wanted to share this little "thing" I came up with while plunking on the guitar, in the living room. I was asking myself why I didn't keep my mouth shut when I was talking to Chad Cunningham.

Surprisingly I passed the audition and was invited to learn a few of the bands songs and play on stage with them at an upcoming outdoor concert at Wenona Park, on the Saginaw River. The chemistry, with the guys, was right and this green horn guitar picker wound up playing and recording with the Robert Lee Blues Band for a couple of years, with "Bluest Blues" still one of my favorite songs.

So what does strapping on an electric guitar have to do with this account of restoring an historic home, the woman who once lived there, or the three Great Lakes captains? This is where opportunity again began knocking, although I didn't know it at the time. Through trial, error and tribulation, this writer, has learned again and again, that opportunity often knocks very softly, and one must listen very closely, to find out exactly where that door is.

One night while playing with the Robert Lee Blues Band at O'Hare's nightclub in Bay City, Michigan, the faint knock of opportunity would have been heard, if it were not for the bands blaring amplifiers. O'Hare's once belonged to Emil Westover, a saloon keeper; it is where Captain William Neal was said to have had a snort on occasion, and the same saloon where Captain William Neal's friends toasted his sudden passing in 1896. It was also at O'Hare's, known at the time as Larson and Rayman's, where Captain William Neal's son, Captain Walter Neal, the "Iron Man," was known to have bent his elbow a time or two. Uncle Walter only lived a few short blocks from the watering hole, so it was convenient. Opportunity was knocking, and the Robert Lee Blues Band was rocking, and little did I know, that yet another piece of this story, would soon find me.

During the course of the evening at O'Hare's, the band invited patrons to come up on stage, to either sing, or have a little fun by playing the tambourine with the band. I always enjoyed that part of the evening simply because it was a lot of fun and the customers seemed to enjoy it as well, as they jammed with the band. This time opportunity came in the form of the jingle jangle of the tambourine. It was one of those guest tambourine players who had some of the answers to more than a few of my questions about 'Uncle Walter' Neal, but of course, I didn't know it at the time.

Combing through death records, cemetery records and obituaries I came across a couple of names, that led me to the great-granddaughter of the "Iron Man," who luckily, still lived in the area. After a few telephone calls, I made contact with Anne Sullivan, great-granddaughter of Captain Walter Neal. After introducing myself on the phone, and the reason for the call, I asked if we could meet so she could share her family stories about the good captain.

Anne said, "Well golly, we've already met, I was the "Anne" who was playing the tambourine right next to you, with your band at O'Hare's a couple of weeks ago!" As if on cue, we both began to laugh out loud, what a coincidence, and right under my nose. After all the work in attempting to track down Uncle Walter's great-granddaughter and here she was jamming with the band, at the very spot where her great-grandfather and her great-great-grandfather were once known as regulars during the winter season.

Anne was a great help. While many of her great-grandfather's things had been donated to various museums, she still had some things tucked away, that her mother had given her, before she passed away. The latest discovery, was a large envelope of newspaper clippings that great-grandpa Walter Neal had clipped himself, along with several Neal family photographs that I had never seen before. But Anne had never seen any photos of her great-great-grandfather William Neal or his wife Mary Benton Neal. I was both amused and honored that I would be the one to acquaint Anne with her ancestors; the look on her face and her wide eyes were worth a million.

"Allow me to introduce you to your great-great-grandparents," I said. I laid out the photos of both William and Mary Neal for her, and sat back and watched her eyes, as they studied the faces of her ancestors. What a personal joy it was to connect "Tambourine Anne" to her distant past. And what a treat it was to hear the stories she remembered about her great-grandfather, Uncle Walter. Her recollections are contained within the pages of this book. But that is not the end of Tambourine Anne's involvement or assistance on the project. While studying the ancient photos Anne had never seen before, I had asked if she had anything else, tucked away, that might help tell the story of her family. She had brought out several jewelry boxes that once belonged to her deceased mother Mary, Walter Neal's granddaughter. We tried to match the jewelry depicted in the photos of her grandparents, her great-grandparents and her great-great-grandparents, to the things that were in her late mother's jewelry boxes, but we had no luck.

It had been a wonderful and most fruitful visit with Captain Neal's great-granddaughter, we both laughed about the odds of meeting up on stage, in a night club, where her ancestors had been patrons so many years before. I gathered up the old news paper clippings from the kitchen table and was ready to head for home, to absorb the information in my latest find. I shook Anne's hand a second time and was halfway out the door when a strange look came over her face. Anne paused and said, "Wait a minute there is one more box with some costume jewelry and some other junk we should probably look at before you leave, if you have the time." While Anne went to the back of the house to dig up that last box, I was getting that funny feeling, that if there was anything left to be found, now was the time, or at least I had hoped it was. When she returned to the kitchen, she held up a beat up old yellow jewelry box that had certainly seen better days. The corners of the little treasure chest were broken and the top was held by a single hinge and a broken clasp on the front. Uncle Walter's great-granddaughter put the battered jewelry box on the table, and stepped back away from it. She said, "Do you want to open it, or would you like me to?"

I nodded to Anne to go ahead, we both stood there for a moment or two, just looking at this crumpled old box, then at each other, and then back at the box. Finally Anne opened it, and right there on the top of the heap, was exactly what we had been looking for, her great-great-grandmother's skeleton key pin. "Yeeoow, it's the same pin your great-great-grandmother is wearing in this photo taken in the 1880's and look here, it's the same pin that her daughter Nettie Neal is wearing in the photo taken when she was seventeen years old!" The time lost pin was made of copper and brass, and looked as if Mary Neal or daughter Nettie had just taken it off, it still worked and was in remarkable condition for its age. It was a very exciting moment. Tambourine Anne also held another family secret, that had only been shared around the family's kitchen table. And the secret would have remained so, had the question, not been asked. "Anne, do you have any idea, or have you heard any stories, how your great-grandfather, was able to hang on to the floating wreckage of the *Myron's* pilothouse?" Anne knew all about Captain Neal's 20 hour struggle to hang on, as he was beaten up by the pounding waves. "Well," she said, "my mother told me it's not something my great-grandfather wanted anyone to talk about. But, from what I remember, he was able to hang on the to pilothouse, because the leather coat he was wearing, that he got from my grandfather, would sometimes freeze to the wreckage, and that helped him to hang on, and thank God, or he might have never made it home. That was the part of the story my mother said, he always left out."

Epilogue

Ideas and Other Thoughts

The series of events, written on the preceding pages of this project, would have been impossible without the help and cooperation of Sis, Anna Mae, Anne, and all of the others who let me in the door, gave of their time, and cared enough to point this writer in the right direction, during this most enjoyable journey. However, this effort is by no means the complete and unabridged story of Uncle Walter, Captain Johnston, the Great Lakes, the men who sailed them or their wives.

Contemporary Great Lakes maritime authors like Chris Kohl, Paul Hancock, Glen Shaw, Wes Oleszewski, Frederick Stonehouse, Robert H. Smith, and a long list of others have preserved a more technical side of our maritime history, to learn from and enjoy. It is the writer's intent, to encourage everyone to take a second look at the people who surround us, whether friends, neighbors or even total strangers who may hold precious memories that need to be recorded.

Museums, libraries, and genealogical groups are excellent sources, to help guide new historical researchers, where to look for information, and how to preserve the articles and artifacts that are discovered along the way.

This writer is both humbled and grateful, to those people, who have invested their time and energy in making "The Captain's Chair" project a reality. The search for who we were and where our ancestors have taken us isn't over yet. Only but a few of the many questions, about the past, have been answered here, but it is a beginning, for further exploration.

Thank you Sis Johnston, for taking the chain off the door, for your patience, and the gift of your precious time. Without you, none of this would have been at all possible. But your china doll, has not yet been found. Sis made sure the story of her family and Uncle Walter was accurate, she also assisted in the grammatical application of the material as well. Sis had asked that the entire draft be read to her, once the project was completed.

After hearing the entire story, as she rested in her easy chair, she nodded her head at its conclusion and said, "nice job." Then she asked when she would have her own signed copy of the book. Two days after the reading, she took her last breath and passed away. It was just like her, not to leave any chore undone.

Many wonderful discoveries have been made on this journey, and many lessons learned as well. We are much richer, as people, and as benefactors of Sis Johnston's time treasured recollections, and her gifts to us, of the Captain's Chair, and his Charm. Meanwhile, the search continues for Sis's missing China Doll.

Captain Alexander Johnston's daughter learned exactly what her father had taught her, discipline, tenacity, and the ability to dead reckon during a gale of any sort. She was tough with her students only because she cared about them. She was rigid when order demanded it, but she was also a gentle and gracious lady by nature.

Alexandra Stowe Johnston, after a full life, passed away just before 8:00 o'clock, on the morning of May 1st, 2003. Sis lived long enough to read this entire book and help in the selection of the photographs used in these pages.

Alexandra "Sis" Johnston with Jay Brandow /
photo Glen Groeschen

On the wall next to her bed was the latest photograph of the house she and her family once knew as home. Sis pointed to the photograph and winked, it was her way of saying, "I like that."

Captain Johnston's old house now has more than a new roof; it has its history, it has its identity, and with any luck, will remain standing for generations yet to come.

Through the works of those authors who have made the painstaking effort to research and track down information; we as readers can enjoy in a few hours what it may have taken the writer years to assemble. They share their passion for the past, newfound discoveries and their concern for history's preservation. Through printed words, photographs and the music our ancestors; their stories are still very much alive. The following roster of artists offers a variety of options to please all of the senses.

For the underwater adventurer, there are well-written reference guides that will locate a shipwreck; and caution scuba divers on particular conditions should one choose to explore the protected underwater preserves of the Great Lakes graveyards. Other works will afford the armchair scuba diver all of the action, packed between the covers of their books, to be enjoyed, nestled in a comfortable chair.

Over recent years, video taped interviews have captured the fading images of those who helped to write or were witness to the events chronicled as history. While words and pictures can bring us closer to the past, music from the bygone era can whisk us back in time through song. The work of the followings artists are but a few ideas for the curious.

Suggested Reading

Boats, Great Lakes And Me by Capt. Glen Shaw
Recipe For A Community by Patricia Drury & Ron Bloomfield
On The Banks of the Beautiful Saugenah (a series) by Roselynn Ederer
Shipwrecks of the Great Lakes by Paul Hancock
Bay City in Postcards by Leon Katzinger
Michigan's Lumbertowns 1870-1905 by Jeremy W. Kilar
The One Hundred Best Great Lakes Shipwrecks by Cris Kohl
Mysteries and Histories by Wes Oleszewski
Went Missing by Frederick Stonehouse
Paul Bunyan; How a Terrible Timber Fella Began a Legend by D. Lawrence Rogers
The Long Ships Passing by Walter Havighurst
Vessels Built on the Saginaw by Swayze, Roberts & Comtois
Great Lakes Circle Tour by Bob & Ginger Schmidt
Voices of the Lakes by Steve Harrington
Maritime Museums of North America by Robert H. Smith

Videos:

Graveyard Of The Great Lakes / Great Lakes Shipwreck Historical Society
Great Lakes in Depth / *The PBS Series* / Ric Mixter Airworthy Productions, Saginaw, Michigan. www.Lakefury.com

Music:

Dougie Maclean / Ready for the Storm / Dunkeld Records, Scotland. Tel. 44-011-(01350) 727686 (from the US) / *www.dougiemaclean.com*

Robert Lee Blues Band / www.robertlee52.tripod.com

Hoolie / Great Lakes Folk Lore / jjc53@chartermi.net

Dan Hall / It's Quiet Where They Sleep / www.danhall.com

If the books, videos and the music seem only to be an appetizer, it is still possible to wrap ones self even further in the elegance and grandeur of the Victorian Era by booking a suite in a Great Lakes area Bed and Breakfast. The list is too lengthy to publish here but many wonderful operations can be found along the shore towns of all five Lakes. In most cases, logging on to the Internet and searching the names of port cities will yield a wide selection of historic lodging options, historic landmarks, museums and cities full of people who are proud of the heritage. If by chance one should find themselves touring the sites mentioned in the Bay City/Saginaw/Saginaw Bay area, there are two Grand Victorian era homes, with their own wonderful histories, that now serve as Bed and Breakfast operations. They are located across the street from each other, and the owners often work together to ensure the comfort of each other's guests.

Bed and Breakfast Operations:

The Clements Inn, [ClementsInn.com] Queen Anne Shingle Style (built in 1886) Elegant décor, detailed with Oak wood paneled walls and ceilings.

Keswick Manor, [KeswickManor.com] Elegant Georgian Greek Revival (built in 1896)

For the vacationers who wish to explore the shoreline of the Great Lakes, there are countless rustic cabins and modern hotel/motel operations to accommodate your wondering crew. For those who wish to touch history and go beyond the books, music, videos and the historic homes there are several locations across the lakes where you can board a vessel and feel the spray of the sea on your face. The Traverse City Area is home to a tall ship that will take passengers for an excursion, and there is also a boat service that offers a dinner cruise on a modern pleasure vessel located in the area of Charlevoix, Mi.

Glass bottom boat tours are available in the Munising, Michigan, area where several wrecks are nearly perfectly preserved and can be appreciated by non-divers.

Bay City, Michigan, is home to two steel-hulled tall ships, *The Appledore IV*, and the *Appledore V*; both are used as floating classrooms for environmental studies that serve students of all ages under her sails, from across the region. During non-class time hours both vessels serve as tour ships to the public with excursions up the Saginaw River to the City of Saginaw or out into the Saginaw Bay. Those who would like to learn more about the BaySail/Appledore Education Program should contact the Bay County Chamber of Commerce in Bay City, Michigan, USA, for more information. *The Princess Wenonah*, a converted automobile ferry, now serves as a passenger tour boat and a floating hall for rent, that traverses the Saginaw River and the bay on a regular basis. The daily excursions on the *Princess* have traditionally been underwritten by area merchants, allowing several hundred passengers at a time to enjoy the 3-hour courtesy.

If preserving a part of Victorian history appeals; the preservation of the Chapel at Elm Lawn Cemetery is being spearheaded by the "Friends of Elm Lawn Chapel." The group has obtained non-profit status, and is seeking help from public and private donations to restore the structure to its original Victorian grandeur. Interested parties can contact chapel restoration organizers at, *Elm Lawn Cemetery, 300 Ridge Road, Bay City, Michigan*, USA, *48708*. The aforementioned listings are by no means a total or complete compilation of materials or activities available in the region. A more complete list of Great Lakes Maritime history materials, museums, research sites and other fun spots, including the best locations to watch modern day Great Lakes ship traffic can be found by logging on to BOATNERD.com The site offers numerous links to nearly every facet of the lakes, from the people, to the ships, and to the many museums and universities that keep record of them, salties too. Enjoy the journey.

Bibliography

Bay County Past and Present 1918 & 1957
Bay City (Polk) Directory 1872/73, 1896, 1907, 1919
History of the Great Lakes / Volume II (page 490)
Michigan's Lumber Towns, 1870-1905, Dr. Jeremy W. Kilar
The Broadcast Century & Beyond, Hilliard & Keith
Portrait and Biographical Record of Saginaw and Bay
    Counties. / 1892
West Bay City Tribune, Dec. 6, 8, 9, 1896
The Bay City Tribune, Dec. 30, 1897
Bay City Times Tribune Nov. 23, 25, 26, 27, 28, 29, 1919
    Dec. 5, 1919
Bay City Times, May 14, 1921
Bay City Times, Sept. 23, 1928
Bay City Times, Jan. 9, 1941
Bay City Times, April 29, 1942
Bay City Times, March 21, 1951
Buffalo, New York Newspaper 1938
News Chronicle/ Port Arthur, Ontario, Canada Nov. 24, 25,
    26, 27, 28, 1919
The Detroit News, August 16, 17, 1935

A very special thank you to:

Maritime Photographs, Courtesy of Ralph Roberts

The Bay County Historical Society, Bay City, Michigan.
Executive Director: Gay McInerney

Curator of Collections & Research / Ron Bloomfield /
Historical Museum of Bay County, Michigan

Architectural Historian / Dale Wolicki / Historical Museum
of Bay County, Michigan

The Great Lakes Shipwreck Historical Society, Sault Ste.
Marie, Michigan.
Executive Director: Tom Farnquist / Development Officer:
Sean Ley

The Saginaw River Marine Historical Society, Bay City,
Michigan
Executive Director: Don Comtois / Vice President: Don Morin

The Great Lakes Lore Museum, Rogers City, Michigan
Director: Ed Brklacich

The Saginaw County Historical Society /
Saginaw County Historical Museum

The Bay County Public Library System and its crew.

Clarke Museum, on the campus of Central Michigan
University, Mt. Pleasant, Mi.

Anna Mae & Larry Wires / Bay City, Michigan

Elm Lawn Cemetery / Bay County, Michigan

Leon Katzinger / Bay City, Michigan

Special thanks to: Marc and Mar y Jo Scott, Glen
Groeschen, Eric Jylha, Dick VanNostrand, Pam Taber, Kai
Rolann, Tom Nolan, Tom Knaub, Captain Dan Arsenault,
Cheryl Harvey, Kyle Higgs, William Gildenstern, Les Root,
Garner & Sandy Train, Vickie Stuart, Mike VanHorn, Roy
Walton, Terry Watson, Sean Waterman, Dr. Devere Woods
Jr. PhD., Kurt Bauer, Paul Gradowski, Dave & Sharron
Hetzel, Captain Al Flood, Cornelius M. Lutze, Chuck & Bea
Chapin, Dick Kernan, Lucy Norman, Carlton & Andrea
Coffey, Edju Wisniewski, Leonard Norman, Jim Osterman,
Nancy Vader-McCormick, Brian Donaldson, Elaine & Bill
Fournier, Bill Jackson, Marlin Kibbe, Tom Majchrzak, Linda
Theaker, Jill Wellington, Frank Bryden Jr., Deb Lang, The
City of Bay City, Mi., The O'Hare's crew, Mark & Mary
Sayles, Kay Cole, Dr. Dennis Couture, Mark & Punk Lungren,
Mark Zahnow, Dick Zacharko, Jim & Victoria Hill, The Tuthill
Family, The Robert Lee Blues Band Road Crew, Vince Stuart,
Matt, Bill, Luke and Tim, Lagalo & Family, Mary & Emil
Rivard, Linda Plackowski, Timothy Burns, B.J. Jones, Mary
& Chuck Militello, Gary Linkowski, Donna Lowe, Ray Tudor,
Captain "Chuck" Muddy Waters, John DeJohn, George &

Joan Croft, Larry Butcher, Derrick Smith, Jack Frasik, Ellen B. Waxman, Tim O'Neal, David Guard, Jerry Casault, Tim Hall, Erik Horn, Mary Vogelaar, William Anderson, Laura Dull, Tim Brandow, Wilbur & Bud Brandow, C. Brandow, L. Brandow, and to Lula; whose vision and inspiration made it all possible.